THE

SHIP CANAL

TO CONNECT THE

ATLANTIC & PACIFIC OCEANS.

WITH A HISTORY OF THE ENTERPRISE

From its first Inception to the Completion of the Surveys.

INCLUDING THE INSTRUCTIONS FROM F. M. KELLEY, ESQ., TO WILLIAM KENNISH, ESQ., CIVIL ENGINEER—REPORT OF MR. KENNISH'S SURVEY, WITH ACCOMPANYING PLATES, AND A PAPER UPON THE THEORY OF THE TIDES—CONFIRMATORY REPORT OF E. W. SERRELL, ESQ., CONSULTING ENGINEER—AND AN ESSAY UPON THE IMPORTANCE OF THE CANAL IN ITS RELATIONS TO THE COMMERCE OF THE WORLD.

NEW-YORK:
GEORGE F. NESBITT & CO., PRINTERS AND STATIONERS,
Corner Pearl and Pine Streets.

1855.

REPORTS, ETC.

INSTRUCTIONS TO CIVIL ENGINEER.

NEW-YORK, *November 2d*, 1854.

WILLIAM KENNISH, ESQ., *Civil Engineer.*

DEAR SIR:

Be pleased to form a party forthwith for the purpose of making explorations for a ship canal in New Granada, and proceed at once to Panama, *via.* Aspinwall, by the next regular steamer.

At Panama charter such a vessel as, in your judgment, will best suit the purpose, and follow down the coast to the southward, with a view of discovering a good harbor in the vicinity of latitude 7 deg. north.

Having found such a bay as will admit the largest class vessels, make a regular hydrographical survey of the same, noticing in particular such improvements (if any are required) that can be made, to render it perfectly safe under all circumstances.

While passing down the coast, observe closely the mountain ranges, and look for any decided breaking down of the Cordilleras that occurs.

From the bay (if you are fortunate enough to find one), proceed eastward towards the dividing ridge of the country, and search for such a line as will admit the construction of an open cut, without locks, for a ship canal that shall connect the waters of the Pacific Ocean with the Atrato River, near its confluence with the Truando River, in latitude 7 deg., longitude 77 deg. west from Greenwich.

The researches which have heretofore been made at my instance, and those made by distinguished travelers, indicate that the summit of the country in this general direction is very low; and in order to establish the lowest pass—having reached the ridge between the waters flowing into the Pacific and those discharging into the Atrato River in an opposite direction—you will run an instrumental line of levels along the summit of this ridge, in a direction transverse to your general line.

By so doing, the lowest point will be determined, and into this gap, from the Pacific, direct your researches. It is important, when the general direction is approximately determined, that a regular transit line should be run, and carefully measured, and all the inequalities of the surface ascertained by level, in the usual manner.

When you have crossed the dividing ridge in the best place, proceed to the Atrato River, on the most practicable route, and then descend to the Atlantic Ocean, continuing your examinations of direction, currents, soundings, &c., &c.

At the mouth of the Atrato make the necessary examinations for harbors, &c.

The more immediate object of these surveys being to determine the approximate cost of an open cut, without locks, from ocean to ocean, having thirty feet depth of water at extreme low tide, and sufficiently wide to pass two of the largest steamers afloat, you will fill up the details of the examination with reference to this purpose.

All particulars that your time and circumstances will permit you to record, of climate, and the natural productions of all kinds, of the country, be pleased to note.

The sanitary condition of this region requires your special attention.

As it is of the greatest consequence that whatever is done should be most thorough, please provide yourself with all instruments and apparatus that will ensure exactness.

Report to me by letter, from time to time, as opportunity offers, and when you have finished, please return to New-York as expeditiously as possible.

Herewith I hand you such letters of credit and introduction as are necessary, and will facilitate you in the prosecution of the enterprise.

Wishing you and your party health, and a safe return, and confiding in your ability, energy, and faithfulness,

I remain, yours truly,

F. M. KELLEY.

REPORT OF CIVIL ENGINEER.

F. M. KELLEY, Esq.

SIR:

In compliance with your order, bearing date November 2d, 1854, directing me to proceed to the Province of Choco, in the Republic of New Granada, South America, and explore a route across the Cordilleras, from 7 deg. north latitude, on the Pacific, to 7 deg. north latitude, 77 deg. longitude west of Greenwich, on the Atrato River, for the purpose of locating a line for an *inter-oceanic ship canal*, without locks, I have the honor to

REPORT:

That, accompanied by my first assistant, Mr. Norman Rude, I started, in the steamship George Law, for Aspinwall, on the sixth November, 1854, and arrived there on the fourteenth day of the same month, whence we proceeded across the Isthmus, and arrived at Panama the next night, where I was joined by my second assistant, Dr. R. G. Jameson.

December 10, 1854.—We sailed from Panama for the Island of Tobago, and arrived there the next day. Here we remained, in order to get necessary alterations and repairs made to a bungo, which we purchased for the voyage down the coast.

December 13.—Sailed at one o'clock, A. M., for the Pearl Islands, which were in our course; the wind light.

December 14.—We laid to all night, and reached the pueblo of San Miguel, on the Island San Miguel (one of the Pearl group); the weather clear and pleasant.

December 15.—At San Miguel. The 16th, 17th and 18th were occupied in the voyage from the Pearl Islands to the Boca Chica, the smallest of the two mouths that connect the harbor of Darien with the Gulf.

We stopped the night of the 18th at Palma, a pueblo, or village, of five houses, built of cane, and situated just inside the harbor of Darien.

December 19.—Arrived at two o'clock, P. M., at Chapigana, and landed at the residence of Messrs. Hossack and Nelson.

Here the Alcalde and village Judge called upon us, and offered their services, and gave us information of a very encouraging nature respecting the country between the Jurador River and the river Atrato.

They expressed much apprehension of being attacked by their neighbors, the San Blus Indians, who, ever since the memorable and unfortunate expedition of the Virago, and of the American party, under Lieut. Strain, have been more inveterate than before against foreigners.

The object which we had in view in visiting this place, was to obtain the aid of Mr. Nelson, and one or two sailors acquainted with that portion of the Pacific coast which extends from the Gulf of San Miguel to the river Jurador. Skillful and experienced pilots are very necessary for this voyage, as the coast is bold and rocky, with numerous reefs and strong currents. The thermometer was 82 deg. at noon.

December 22.—We obtained two sailors, one of whom engaged for the voyage; the other would not go further than Garachine. Mr. Nelson also agreed to accompany us. Here our cook deserted, but was retaken, and sent on board by the Alcalde.

We returned from Chapigana to Palma, which we reached in the afternoon, and found the tidal currents of the harbor of San Miguel very strong.

December 23.—Left Palma for Garachine, passing the Boca Chica; emerging from this narrow and remarkable passage, we steered for Garachine, and reached it about 5.30 o'clock, P. M.

Here we found that it would be impossible to obtain the services of any sailors until after Christmas, and were therefore constrained to remain until the 27th.

December 24.—We took altitudes, &c., in order to determine the longitude of Garachine.

December 27.—Obtained a pilot; also another to replace the one who left us at Garachine.

At Garachine we noticed a lofty range of mountains, one of which, immediately behind the village, rises to the height of 3,000 feet; and from this the range passes southwards, following the coast to Puerto Pinas, where it is but little diminished in altitude.

December 28.—We made sail and rounded Cape Garachine, and being apprehensive of not having water enough on board, we put in at Puerto Escondido, a small boat harbor, very inaccessible, with narrow entrance, through which there rushes a furious tidal current, causing a heavy swell and formidable surf.

When safely anchored inside, we obtained excellent water; and suspending our hammocks from the branches of trees, passed a comfortable night.

December 29.—Sailed again; the wind being light and baffling, we were compelled to anchor during the night.

There are few situations along the Pacific, in this latitude, where good anchorage is obtainable. The coast all the way from Garachine, is rocky and bold, and has but little beach.

The hills, from the margin of the sea to the highest visible mountain ridges, are covered with dense forests, rarely traversed by any human being. This is true of the coast all the way from Punto Garachine to Puerto Pinas, and from thence to Punto Ardita.

With the exception of Puerto Pinas, there is no harbor of sufficient depth for vessels larger than those used for coasting, (which are small,) on all this line of coast. Puerto Pinas, however, is capacious, and deserves a more extended description, which is given hereafter.

December 30.—We arrived within a mile and a-half of the entrance of Puerto Pinas, but were baffled by the winds, and were compelled to anchor in a dangerous situation, exposed to a heavy swell, which threatened to drive us ashore.

December 31.—We weighed anchor at daylight, and, after four hours' heavy pulling against a current, made the entrance of Puerto Pinas, near which there are several remarkable detached rocks, which mark the position of the harbor very distinctly.

We entered the harbor about eight o'clock, A. M., and followed closely the rocks on its northern margin, until we reached the mouth of a river which flows into it from the direction of the Sierra, which is lofty, and limits the view towards the interior.

At the time of landing, the thermometer stood at 88 deg. A rancho was speedily erected on the sandy neck, between the river and the bay.

This harbor is two miles and a-half wide at the mouth, and extends inwards, to the crown of the bay, about five miles. It is closely hemmed in by mountains, which are densely wooded, and rise to the altitude of from five hundred to one thousand feet.

The more distant ranges, in the interior, appear to have an altitude from three thousand to four thousand feet. The shores, on both sides, are indented with bays, which would shelter a vessel against all winds. These are of limited extent, but have deep water close to the shore.

The Rio de Pinas is an inconsiderable stream, liable to sudden and heavy freshets, during one of which, our bungo dragged her anchor, and was with great difficulty prevented from drifting out into the bay.

The playa, or beach, which margins the crown of the harbor, is very smooth, with a gradual slope, but, owing to the tidal current, is at all times beaten by a succession of breakers, rendering it difficult, and even dangerous, to land upon from boats. Landing-places can only be found at the mouth of the river, or in the various inlets which indent its northern and southern shores.

January 1, 1855.—I started from the rancho at half-past eight o'clock, A. M., and, accompanied by the Doctor, walked towards the crown of the harbor, intending to ascend one of the hills. We were attended by our pilot and the sailors, armed with hatchets, to cut the brushwood in our ascent.

After walking three miles along the beach, we reached a cool and shaded rivulet, falling in a cascade. Here we breakfasted, and then commenced the ascent.

We took with us one aneroid barometer and detached thermometer; the other aneroid, with thermometer, being left at the rancho, with Captain Rude, who was directed to note, every hour, the barometer pressure, and register the atmospheric temperature.

When we reached the summit of the hill, which was very steep, and difficult of ascent, and, moreover, extremely slippery from the clayey nature of the surface, we observed the instruments, and found the atmospheric pressure diminished 45-100 of an inch, indicating, with the correction for the temperature, a height of nearly 500 feet.

January 2.—We took angles of elevation, to ascertain, trigonometrically, the height of the hill which we ascended yesterday; but the angles obtained were, necessarily, very acute, and consequently, unsatisfactory.

In ascending and measuring the altitude of this hill, we had two objects in view: First, to test the accuracy of a barometrical observation, as compared with a trigonometrical one; and, secondly, to obtain a view, from an elevated summit, over the neighboring country. But in the latter we were completely disappointed; the foliage being so dense, that it was impossible to see to a greater distance than a few yards.

This difficulty we had to contend with during all our subsequent travels, from the Pacific to the Atrato. Hence, all statements respecting views from the tops of hills in New Granada, are to be received with caution and distrust. No extensive prospects are obtainable from any mountain seen by our expedition in New Granada.

January 3.—At Puerto Pinas, we became acquainted with a resident of Jurador, who gave us a very intelligent account of the country between the mouth of that river and the Atrato.

January 6.—This evening, at seven o'clock, we got under way, and stood out of the harbor, with light, unsteady winds. We passed a row of remarkable detached rocks, seven in number, which mark the southern head of the harbor.

January 7.—We reached Punto Ardita, after two days of difficult, harassing, and tedious navigation; making sail when the wind was favorable, and anchoring when it was calm, or foul. The coast between Puerto Pinas and Punto Ardita, is bold, rocky, and dangerous. There are two remarkable promontories. The most northerly is called Punto Muerto; the more southerly, Punto Caracoles. In the vicinity of these points, the small coasting canoes and piraguas find anchorage. Punto Cocalito is another similar promontory, within about six miles of Punto Ardita. At this last-named promontory commences the remarkable and important bay, which encloses the outlet of the Jurador and Paracuchichi rivers, besides others of lesser note; and here we perceived, at a glance, that depression of the Cordilleras, (from the altitude of thousands to that of hundreds of feet,) where we have determined to effect a hitherto unexplored route from the Pacific to the Atrato. This important bay terminates, towards the south, in the remarkable promontory of Punto Marzo. An extent of not less than thirty-five miles, with a depth of fifteen miles, from the crown of the bay, to a straight line, connects the two headlands. This line of coast presents three great playas, or beaches of sand; the first of which, forming the segment of a circle, intervenes between Punto Ardita and the mouths of the Jurador; the second runs nearly in a straight line, from the Jurador to the mouths of the Paracuchichi, a distance of ten miles; the third extends from the mouth of the last-named river, in a semi-circular direction, to the mouth of the Corredor River, a distance of fifteen miles.

The waves of the Pacific break in long continuous lines of surf against these elongated beaches, from whence the water deepens very regularly and gradually, affording anchorage of from ten to thirty fathoms, within two or three miles of the shore. The bottom is of sand, deepening very gradually from the shore outwards.

Off the points of Ardita, Jurador, and Marzo, there are detached rocks; but all other parts of the bay are free from these dangerous obstructions.

At Corredor, there is an indentation of the coast, forming a safe anchorage and harbor, where ships of large size can be protected from nearly every wind, within seven or eight miles, running distance, from the mouth of the River Paracuchichi, the importance of which will hereafter be pointed out.

The coast between Punto Ardita and Marzo is deserving of an especial description. First: towards the sea, is a beach skirted with innumerable cocoa-nut trees; more interiorly, a level expanse thickly covered with mangroves, and traversed by canoes or canals formed by nature, in a network, in which the tide rises and falls, showing them to communicate with the Jurador and Paracuchichi rivers.

Beyond this level, or "manglade," the surface rises into gentle elevations, of inconsiderable altitude, the highest of which does not appear to exceed a few hundred feet. Throughout its entire extent, the country, far and near, is covered with trees.

From Corredor to Jurador, and even to Ardita, it is possible to go on foot along the beaches, except at high water, when the tide reaches the bushes. This being the only part of the coast at which a depression of the Cordilleras is visible, the character of the coast line itself requires an especial description, with reference to the existence of a harbor at which a canal might terminate. Hitherto we had not heard of the existence of any natural harbor in this vicinity, except Corredor.

It was necessary for us to enter the mouth of the Paracuchichi, through its rows of breakers, ere we became aware of the wide and beautiful expanse of sheltered water, which was hidden from our sight as we passed along by the long peninsula of Paracuchichi.

To resume the narrative of events: we reached, as already observed, Punto Ardita, and brought up in a small anchorage, only accessible to canoes or bungos. From this point our pilot pointed out the entrance of the Jurador, about five miles distant, and he was of opinion that the surf was too heavy to attempt to enter; nevertheless, about five o'clock, P. M., he changed his mind, and ran across the bay of Ardita to within a few hundred yards of the mouth of the river; here we anchored, as it was getting dark, in about eight fathoms.

January 8.—We got under way at daylight, and pulled close to the northern mouth of the Jurador; but the surf was so heavy and formidable, that to have attempted an entrance would have been madness. The passage is only a few yards wide, and can only be approached at certain times. Formidable rocks break the heavy rollers into foam, which is dashed upwards in clouds of spray.

Steering on for about one mile and a-quarter, we passed the island which separates the two mouths of the Jurador; it is level at each extremity, and rises in its central part to a height of about thirty feet. The surf, as we sailed along, forbade any attempts at landing, and when we reached the southern mouth, the same difficulty met us; in fact, we had no alternative but to steer along the coast from the Jurador in search of a landing-place, but which we could not find nearer than Corredor. There we brought up in a finely sheltered harbor about two o'clock P. M.

January 9.—At Corredor there are no inhabitants. The hills rise abruptly from the beach, and are, as everywhere else in this country, thickly wooded from base to summit. This day we took a line of soundings across the harbor, at a cable's length from the shore, and the result was a medium depth of three fathoms. Leewardly there is a safe anchorage for large ships; protection against the prevailing winds being afforded by the neighboring coasts.

January 10.—We took bearings of important points from Corredor. Mouth of the Paracuchichi N. 25 deg. W.; Cocalito, N. W.; Mouth of Jurador N. W. and by W.

In the evening, having obtained the services of a more trustworthy pilot, who came down from Jurador on purpose to see us, we weighed anchor and stood across the bay, steering for the mouths of the Paracuchichi.

January 11.—This morning, under the pilotage of a skillful seaman, we approached the mouths of the river, from our anchorage off Paracuchichi. The surf was, as usual, high and continuous; there was no passage except through the breakers, and there being no wind, it was necessary to trust solely to the three oarsmen. The pilot, keeping the boat's head at right angles to the line of surf, which afforded our only chance of effecting a safe passage, urged the sailors, by every possible effort of voice and gesture, to pull through vigorously. Three heavy breakers followed us up in succession, and the fourth brought us into smoother water. In a few minutes more we found ourselves floating on the surface of a river, which wound with a slow current through a level space, covered with mangroves of great height. From thence we proceeded in smaller canoes, furnished by a native resident, to the pueblo or village of Eurachichi, which we reached about half-past twelve o'clock, noon, leaving our bungo to be taken after us, when dismasted.

Arrived at the pueblo, which contains about six houses, widely scattered, we beheld with pleasure a fine and spacious inlet, extending northwards from the mouth of the Paracuchichi River, from three and a-half to four miles in length, and from two hundred and fifty to five hundred yards in breadth.

Seen from its southern extremity, it appeared like a fine inland lake, with a perfectly smooth surface, protected from the winds of the Pacific by an intervening peninsula, which supports an abundant vegetation of cocoa-nuts, palm and other tropical plants, growing so densely, that, unless by following a beaten track, it is quite impossible to traverse it.

On the other side of the inlet, the vegetation is of a more sombre aspect, consisting chiefly of manglade or mangrove, which, although seldom attaining great thickness, is distinguished by the extreme hardness and durability of its wood. This wood would, no doubt, be suitable for piles, and may be obtained in any required quantity.

This inlet is not laid down on any chart; it is situated in latitude 6 deg. 57 min. 32 sec. north. We found the temperature cool and agreeable, and heard of no sickness, except a little fever and ague among the natives of the vicinity.

The thermometer ranged from 84 deg. at noon to 70 deg. at night. The barometer from 29 35-100 inches to 29 42-100 inches; determined by the aneroid. The highest tide observed at Paracuchichi, at spring and neap, was twelve feet six inches, and the lowest ten feet eleven inches.

The soundings of this inlet, taken at low water, were from two and a-half to three fathoms, in mid-channel. The bottom is mud and sand, easily excavated to any necessary depth.

Towards the north the inlet narrows to the breadth of twenty yards, and winds intricately. In this direction it can be navigated in canoes for nearly two-thirds of the distance to Jurador; but by degrees the channel becomes narrower and shallower, until it finally loses itself in mud, which prevents any further progress in that direction.

In full states of the tide canoes can advance considerably further, so as to meet a creek that is sent off from the southern channel of the Jurador, thereby converting the Peninsula of Paracuchichi, for the time being, into an island. The breadth of this peninsula varies but slightly throughout its whole length of ten miles; the average distance from the margin of the sea to that of the inlet, being from three hundred to five hundred yards.

Its surface is several feet above the highest reach of the sea. The inhabitants informed us that on no occasion, within their recollection, had it been overflowed.

I saw no drift-wood, or any similar indications of its having been inundated within any recent period. If to these reasons, we add the well ascertained fact, that violent storms are almost unheard of occurrences in this region of the Pacific, we shall readily arrive at this important conclusion, viz.: that this peninsula constitutes a permanent barrier, or breakwater, sheltering from the sea a beautiful and tranquil inlet, almost fashioned by nature to serve as a dock or harbor, of ample extent, and suitable in every respect for the terminus of an inter-oceanic canal.

The surf which I have described is a common feature of the Pacific coast of South America; I have noticed it to prevail everywhere, except in situations sheltered from the westerly winds. It is far more formidable in appearance than in reality, especially in a bottom of sloping sand: a boat with sufficient headway can pass through it with safety at any time at Paracuchichi; but a bungo, such as our own, being too large to be adequately impelled by oars, runs much risk of being swamped.

The line of surf extends outwards, about one hundred yards.

The manner of connecting the inlet with the ocean, by a cut through the peninsula, so as to obviate the inconvenience from the surf, is represented in Fig. 10, Plate IV.

Before entering upon the description of our overland route to the Atrato, I consider it necessary to point out the various important conclusions that are established by the preceding narrative, and observations of the coast from Garachine to the Promontory of Punto Marzo.

First.—That the coast between Punto Garachine and Punto Ardita is everywhere bold and mountainous, and the interior more so, and that it contains only one good harbor, viz.: that of Puerto Pinas, which, however useful and interesting in some respects, is of no importance with reference to an inter-oceanic canal.

Second.—The very remarkable depression of the Cordilleras, which is seen opposite that portion of the coast between Punto Ardita and Punto

Marzo. In this interval the country loses the mountainous character entirely, and assumes the appearance of a gradual rise or slope, with hills of little elevation in the distance. This space is bounded on the north by a range of mountains running in a north-easterly direction from the vicinity of Punto Ardita. On the south it is limited by another lofty range of mountains, following a similar course, and terminating at Punto Marzo.

Third.—Because opposite to this depression of the Cordilleras I have discovered an inlet, which, I believe, is hitherto undescribed, presenting ample extent and perfect shelter, capable of being deepened to any required extent, and of being connected with the ocean so as to constitute a terminus to the canal, and which will need, comparatively, but a small outlay in order to render it complete in every respect.

Fourth.—Because there is safe anchorage within a mile or two, as we proved, of this inlet, and also a valuable harbor of refuge at Corredor, distant only seven miles, in a straight line.

Fifth.—Because severe gales or hurricanes seldom occur on this coast. On this voyage, and on four previous occasions, when in this vicinity, in open canoes, sometimes with the gunnel only a few inches above the water, I never experienced any state of the weather at all approaching the nature of a storm, or what is called a gale of wind; gusts of short duration frequently rush down the valleys that open upon the sea-shore, and these would sometimes endanger the small schooners and canoes used in this navigation, but for the precaution taken of steering with the sheet in hand. In ordinary ship navigation, these temporary gusts, seldom of more than five minutes' duration, would cause no danger, or even inconvenience.

In confirmation of these statements, it is also to be remarked, that the loss of a canoe, or other vessel, on this coast is an extremely rare occurrence: yet the anchor commonly used by the natives is only a large stone, with a cable made of the vejuce, or vine tendrils, so abundant in the forests of New Granada.

Vessels equipped in this wretched manner, and at the same time badly manned, as they generally are, would, doubtless, be lost, stranded upon the rocks, or blown out to sea, in very numerous instances, but for the absolutely stormless character of this section of coast. A gale of wind capable of causing a ship to drag her anchor on this coast, would assuredly be considered a singularly rare occurrence.

Weighing all the preceding considerations, the manifest capabilities of the inlet of Paracuchichi, the existence of good anchorage in the offing, and at Corredor, in its immediate vicinity, rendering this inlet in every respect available as the terminus of a canal; the abundance of available hard wood timber in the neighborhood, the absence or extreme rarity of storms in this portion of the coast, and encouraged, moreover, by all the information we could obtain respecting the character of the country in the interior, I determined to make my arrangements for crossing from this point to the Atrato, by the shortest and lowest route that can be found.

January 14.—Accompanied by Dr. Jameson and Mr. Nelson, I this day started for the Jurador River, intending to confer with certain persons residing there respecting the route across from Paracuchichi to the Atrato, and also, if possible, to obtain peons for the journey.

We were two days absent on this expedition, during which time we examined the two mouths of the river, and also noticed the windings and other features of the river itself. The disposition of the mouths of the Jurador is laid down in my remarks on the facilities for rendering them suitable for a canal terminus.

Having ascertained that several days must elapse before the necessary number of peons could be procured at Jurador, but placing full confidence in the activity and intelligence of the individual employed to effect the arrangements on our behalf, we were content to wait patiently until he should report himself and his people to be in readiness for departure.

January 21.—We left our rancho at Paracuchichi this morning at half-past ten o'clock. I was accompanied on the occasion by the Doctor and Mr. Nelson, with our guide, Alexandro, and peons to paddle the two canoes, and carry the baggage on the route.

I left Captain Rude at Paracuchichi, with directions to continue his barometrical observations at certain intervals, and to watch over the safety of the heavier and more valuable baggage, including money and provisions, with which he was directed to follow as soon as we should report the possibility of crossing to the Atrato by this route.

To one or two of the most trustworthy peons was confided the duty of carrying the surveying instruments.

We crossed the inlet in a south-easterly direction, steering for the mouth of a small river, which falls into it about half a mile from the nearest mouth of the Paracuchichi. We found the mouth obstructed by a bar of sand; interiorly of the bar it assumed the character of a sluggish and tortuous stream, about twenty yards in width, and closely hemmed in by the forests of mangrove trees, which generally cover the level spaces, close to the sea-shore, on this coast. A few fallen trees retarded our progress. The banks of the creek on each side consist of a black alluvial soil, easy to cut through.

The highest tides reach to two miles from the mouth of the river; here the banks consist of alluvium, but being less subject to overflows, present a different vegetation—the mangrove being entirely absent. Large spaces are covered with plantain walks, the fruit of which constitutes the main dependence of the native population; it grows with unbounded profusion.

From this point the river becomes quite shallow, with a bottom of sand and pebbles; the latter consisting chiefly of the rolled fragments of clay slate. Six small rapids occur in this part, varying in height from six to eighteen inches, and estimated to amount to six feet in the aggregate.

At the distance of a mile and a-quarter above the tidal reach, we entered the mouth of a tributary called Pie de Nerqua, on the bank of which, about fifty yards from the confluence, we halted at three o'clock, P. M., for the night, and erected No. 1 Rancho, or station.

It had now commenced to rain heavily, and continued without intermission throughout the afternoon and night, causing a rise of only a few inches in the river. The general course from Paracuchichi to No. 1 Rancho, is north-east.

January 22.—Rancho No. 1 being situated on a point or tongue of land, formed by a bend of the Pie de Nerqua, I measured, by level, the height of the river at opposite points, above and below, and found a difference of two feet six inches (2 ft. 6in.) The run of the current I found to be eighteen yards in fourteen seconds in the small rapid, and four yards in fourteen seconds in a piece of still water contiguous.

We then crossed, in succession, three other tongues of land, inclosed by the bends of the river between the crossings; so that our general course lay, as before, north-east rather easterly.

The necks of land were from ten to twenty feet above the level of the river, and consisted entirely of clay. Only at one or two points did the indurated rock present itself, and then rose no higher than the bed of the stream.

The depth of the river was about one foot, so that we had no difficulty in wading through it. Two other measurements of the flow of the current gave, respectively, nine yards in fourteen seconds, and four yards in fourteen seconds. The distance measured in this part of our course was one thousand yards, when we reached the confluence of a small tributary of the Pie de Nerqua, at a spot called by the natives Dos Bocas.

Here we left the Pie de Nerqua, and followed the ridge of a hill which is inclosed between the two rivers, still steering north-east. The ascent of the hill was gradual, with occasional depressions; and on each side, separated only by a breadth varying from twenty to fifty yards, was a deep ravine.

The detached thermometer stood 84 deg. at 11 o'clock, A. M. Our distance from the Dos Bocas was one hundred and thirty-eight yards. We then found the ridge to descend from this point, until it suddenly dipped down to the level of a small, sluggish quebrada or brook, which crossed our road, and marked the termination of the first of a series of remarkable elongated hills, moderate in elevation, and consisting of clay entirely, stretching in a north-east direction, from within a few miles of the Pacific to the valley of the Nerqua.

After crossing this brook, which flowed into the valley on the right-hand side of our route, I ascended another elongated or haystack shaped hill, much lower than the preceding one; and following it, under a heavy rain, still steering north-east, we reached the spot where I determined to halt for the night.

This spot is situated on the bank of the river Chupipi, and is designated in the Field-Book as Rancho No. 2. (See Plates 1 and 3, where the dotted line shows the top of the ridge and where the line of level was run, and the termination of the dotted perpendicular lines shows the bottom of the valley.)

The distance measured from the Dos Bocas was one thousand eight hundred and thirty yards, to the Quebrada, and nine hundred yards from thence to the Chupipi. Here the aneroid barometer, at 1 o'clock, P. M., stood at 29 20-100, and the thermometer at 79 deg.

The Chupipi will cross the line of the canal, as it flows from a northerly direction and passes away to the south-westward to join the Paracuchichi.

From its small size, it will form no impediment to the construction, but will be useful as a feeder. Its average breadth is about twenty feet, with a medium depth not exceeding two feet. Its bottom is pebbly, and here and there a stratified rock of clay slate presents itself in the bed, dipping at an angle of 20 deg. towards the north-east: the banks, for a considerable space above the level of the water, and supporting a heavy vegetation, with trees, from sixty to seventy feet high.

These marginal spaces of alluvial soil are frequent along the rivers of this country, and are of great importance to a public work requiring large bodies of men, from the immense profusion with which they yield the plantain, and other kinds of food.

January 23.—We remained at Rancho No. 2, awaiting communications from Captain Rude. On the same day, the party at the rancho at Chupipi started at seven o'clock, A. M., the thermometer at 80 deg., and, passing over a level space of alluvial soil, began the ascent of another hill, of the same form of the others we had previously passed over; still keeping in a north-east direction, and having close on our right hand a deep valley.

At nine o'clock, A. M., being on the summit of the hill, we observed the aneroid 28 82-100, and the thermometer 79 deg. Here the course becomes east by north, the ridge exceedingly narrow, and we could hear the waters of the Chupipi at the bottom of the ravine on our left, falling over a cascade. At forty-five minutes past ten, a heavy thunder-shower, but of short duration. We continued our journey, alternately ascending and descending, according to the form of the ridge, until a rapid descent brought us to the bank of a quebrada or brook, where, the daily rains having already set in, we erected Rancho No. 3, having passed over, since we left the last rancho in the morning, the distance of two miles and a-half, in the general direction of east north-east.

It again rained very heavily throughout the afternoon and night, accompanied with much thunder and lightning. Aneroid barometer stood at 28 60-100, and the thermometer at 75 deg.

The small quebrada on which Rancho No. 3 was erected, is of no importance, being only three or three and a-half feet wide, with high banks of clay, and a muddy bottom. During very heavy rains it would carry a considerable body of water down to the canal, and would be of some use as a feeder. The hills and ravines continued to be thickly wooded.

I have here to remark again, that in passing successively over the three preceding longitudinal hills of clay foundation, we had invariably observed an uninterrupted valley of great depth on our right-hand side. We look attentively for any transverse ridge or other obstacle to the passage of a canal through this elongated and continuous valley, as it seemed to me, but no such impediment could be discovered.

I have also to point out the important fact, that, although sleeping nightly in the open air, sheltered from the rain only by the sloping roof of palm-leaves, which our peons erected for us in half an hour, our beds consisting of India rubber blankets spread over palm-leaves, none of our party suffered the least in health.

January 25.—We left Rancho No. 3 at nine o'clock, A. M., and commenced the ascent of another elongated hill of the haystack form, consisting, like the preceding ones, of clay in the soft state. At some points, the summit was a mere shell in thickness; in other respects presenting no difference, except in being of much lower elevation: its course is north-east, and the distance to the Chuparador River, which crosses at its north-east foot, is twelve hundred and sixty-five yards.

The Chuparador is a tributary of the Paracuchichi, and is here four feet wide, six or eight inches deep, with banks of clay, and flows south-west, with a very slow current.

Then we gradually ascended another low hill, until we reached a remarkably large tree, (as compared with the others, which are generally of small size on these hills, although of great height.) From thence we steered on about the same course, and crossed another very small brook, or, rather, ditch, falling into the Chuparador. This was the last trace of water flowing into the Pacific, that we crossed.

We then ascended the next in succession of this series of hills. It was of a low elevation, with a table or plateau at the top. On our left hand there was a valley, through which flows the Hingador, a tributary of the Nerqua, and, therefore, the first water hitherto encountered flowing to the Atlantic. The valley of the Hingador is separated from that on our right hand, which is drained into the Pacific, by the hill or elevated ground which we were then traversing. The distance between the two valleys, in some places, is not more than one hundred and thirty feet.

Descending very gradually north-east, we crossed two small branches or tributaries of the Hingador, and reached a large branch of this river, at its junction with one of the small tributaries already mentioned, and erected thereon Rancho No. 4.

In the afternoon the rain fell heavily, and continued until midnight. No rise of the small river before us took place; aneroid barometer 28 6-10.

This branch of the Hingador is a clear, pleasant brook, half a foot in depth, eight feet wide, pebbly bottom, and flows from the westward with a moderate current.

I have to point out that, from the last station or rancho, we carried along with us on our right the same continuous valley; but have to observe that, in approaching the Hingador, the country becomes more intricate, and several ridges proceed from this vicinity as from a centre, in different directions. Two form between them a deep and narrow gorge, through which the Hingador descends to the Nerqua, over a series of water-falls—the first of which I ascertained to be of the height of hundred and sixty feet; the others, seven in number, averaging in height from three feet to twenty-four feet.

This formation of the country will cause a deflection or bend in the line of the canal, in order to bring it to the level of the Nerqua.

The length of the gorge of the Hingador, from the great fall to the Nerqua in a direct line, is not more than one mile and a half. A parallel line from the valley in which it is proposed to bend the canal (the right-hand valley so often mentioned) to near the mouth of the Hingador, would but little exceed that length.

January 26.—Leaving Rancho No. 4 at nine o'clock A. M., we crossed successively two bends of the tributary, and then the main body of the Hingador, which we found to be considerably swollen and muddy from the rain of last night. A fallen tree across the stream served us for a bridge; the current was too strong and deep to ford: the bottom is rocky, and formed by ledges of gray slate. From Rancho No. 3 to this crossing, the distance is sixteen hundred and sixty yards; but as this is not within the limits of the canal, I do not enlarge the details.

The great fall is but a short distance below this crossing, and in favorable states of the atmosphere can be heard easily. From this point we continued our journey along another elongated hill in the plain of the Nerqua, its foot being crossed by a small stony brook leading into that river.

The peons had provided a couple of small canoes, belonging to the only

Indian living on the Nerqua, and in these we descended the river a mile, to the house of this Indian, in whose neighborhood, at a point two-thirds of a mile above the mouth of the Hingador, we erected Rancho No. 5.

Here it was necessary to await the arrival of Captain Rude, from Paracuchichi, with the remaining baggage and provisions, as I had determined to continue our journey from this point to the Atrato, by the Nerqua and Truando.

It was afterwards ascertained that the canoes belonging to the Indian would be much too small to carry the expedition and baggage; it therefore became necessary to have one built of sufficient size for that purpose.

The Nerqua is about the average width of twenty yards. At the fords, which are numerous, it is not more than knee deep; the intervening pools are from three to six feet deep. I ascertained, by measurements, the rate of its flow to be two miles per hour; its water, like that of all the streams crossed on our route, is very sweet and cool, and perfectly salubrious. It flows through a wide alluvial valley, finely timbered and extremely fertile, capable, if cultivated, of supporting a vast population.

On our arrival at the Nerqua, we found the aneroid, at half-past eleven o'clock, A. M., stood at 29 25-100, and the thermometer at 80 deg.

January 31.—Captain Rude and Mr. Nelson, with the remainder of the baggage, arrived this day from the Paracuchichi, having performed the journey across in nine hours' traveling time, although the peons carried full loads. Last night a freshet raised the river three feet and a-half, so that the playa, or pebbly beach, in front of our encampment was overflowed; before morning it had quite subsided.

February 1.—In company with the Doctor, I examined the mouth of the Hingador, and ascended seven hundred yards. It has two mouths, one of which is dry, except in flood; they inclose between them a triangular space of alluvial soil.

February 2.—An Indian, whom we had dispatched to see another Indian living up the Truando, and the only inhabitant on that river above the Nerqua, (or below it, as I afterwards ascertained,) came back, and said that the man we wanted (Juan Domingo) could not come to assist us in descending the Truando, in consequence of his being sick. He returned an answer to our special inquiry, that "no white man had ever been up the Truando, above the falls, or he must have known it."

February 5.—A good tree was obtained for a large canoe, and the Indian and one of his sons, with two of our peons, commenced working on it.

February 7 *and* 8.—We examined the Hingador, from the cataract to the Nerqua, during these days. At two turns below the Tree crossing, the river falls over a horizontal ledge of clay slate, perpendicularly fifteen feet. It then falls at an angle of 45 deg., over large and irregular rocks, of a black and dark color, heavy and indurated, until it reaches the bottom of the descent.

The quantity of water, in ordinary states of the river, is but small; but, when flooded, this cataract must present a most imposing spectacle, from its great height.

As already stated, two spurs, or ridges, form between them the deep and narrow gorge through which this river flows, from the falls to the Nerqua. The walls of the gorge generally rise almost perpendicularly, so that, in following the stream, it is necessary to wade nearly the whole of the way, except where a cataract, or rapid, compels an ascent on the face of the rocks.

The heights, on each side of the river, are from seventy to three hundred feet in perpendicular height. They consist of the black indurated rock, already described. This rock has a ready cleavage, and may be easily excavated.

In only one part of the gorge were observed indications of a land slip, and this had recently taken place. A mass of clay had become detached from near the summit of a hill, on the left hand, and had carried down with it one or two large trees to the bed of the stream.

About one-third of the distance from the great fall, there is another, twenty-four feet high. A little lower down, one of twelve feet, and five others, varying from eight to three feet in height. Besides these cataracts, there are numerous rapids, caused by the descent of the river, and the bottom is strewn with rounded and water-worn masses of rock of all sizes, from that of the smallest pebble, to boulders many tons in weight.

About three-quarters of the distance down the gorge, there are numerous hot springs in the bed of the river, which are highly sulphurous, and raise the temperature of the surrounding water to 110 Fahr.

Below the hot springs, the valley widens out considerably; the river becomes tortuous, forming bends with level tongues of land, in which the vegetation is rank and dense. Finally, as already described, it enters the Nerqua by two mouths, one of which is dry, except during freshets.

February 11.—We measured the Hingador Falls by a line, as follows: —Length, from top to base, two hundred and thirty feet; angle of depression, 45 deg.; resulting height of perpendicular, one hundred and sixty-six feet.

February 15.—We entered a brook, or river, falling into the Nerqua about one-half of a mile below the Hingador, and followed up the same for three hours. Like the Hingador, it passes through a deep and narrow gorge, hemmed in between two ridges. In size, it nearly approaches the Hingador. No cataracts were met with in the bed of the river, as far as we examined it; but there were numerous small rapids.

The distance to which the river was followed up, is computed at one and a-half miles. It is impossible to ascend these rivers, as they are so full of natural obstacles, faster than at the rate of half a mile per hour, on foot, and they are much too shallow even for the smallest-sized canoe.

Another gorge was also examined, opening upon the Nerqua, close to our encampment. It was found to give passage to a small brook, which, at one part, descended in a cataract, ninety feet in height.

Our workmen having reported the canoe ready for service, it appeared, on inspection, adequate in size, and light, being built of cedar, and was, therefore, better for navigating rivers, in which one of the principal obstacles to be contended against consists of fallen trees.

Alexandro Domingo arrived from Jurador and Paracuchichi, and brought with him two young but active Indians, with whose assistance, and that of the two residing on the Nerqua, we proposed descending to the Atrato.

February 16.—At half-past nine o'clock, A. M., we commenced the descent from the mouth of the Hingador towards the Truando.

The Nerqua, below the Hingador, is apparently but little increased, as will be seen by reference to the Table of Stoppages. It presents a great many obstacles, such as shallows and fallen trees.

We passed the following tributaries, which the river receives on the right bank, viz.: at 10h. 30m. the Chupachavi—the one which was examined on the 14th. At 11h. 20m., the Pavarando, which we entered on the left bank; it is a stream rather larger than the Hingador. At 2h. P. M. we passed the mouth of the Equebrador, a tributary of the right bank — equal in size to the Hingador; at 2h. 15m., the Tuniando, a small branch on the left bank.

The banks of the river are alluvial, presenting clay in various stages of induration; generally very soft, but at other times moderately hard, and everywhere much perforated by land crabs; sometimes strata of leaves are observed between thick beds of clay, and there are numerous deposits of a kind of blue clay, which the natives use in making their pottery.

The general course of the river was north-east.

We reached our rancho at four o'clock, P. M. The day, which had been beautiful in the early part of the journey, became rainy towards the afternoon. The barometer stood at 29 2-10; the thermometer at 79 deg.

February 17.—We started from Rancho No. 6 at forty-five minutes past six, A. M. The heavy rains of last night caused a rise of the river of about one foot, which had quite subsided this morning. The bed of the river here consists of shelving clay rock. At ten minutes past seven o'clock the barometer stood at 29 27-100, and the thermometer at 75 deg. There are few or no obstacles in this part of the river; it assumes a slow regular current, with long reaches, the depth being about four feet. At twenty minutes past seven o'clock we passed the mouth of the O'odor, or Oodor, opening on the right bank, with a channel about a quarter the size of the Nerqua, which it enters from about north by west. At thirteen minutes past eight o'clock we reached the confluence of the Nerqua and Truando; the direction of the Nerqua being east by north at the point of junction. The aneroid barometer at this point stood at 29 3-10, and the thermometer at 85 deg.; but during our stoppage here the barometer rose to 29 34-100.

At the confluence we found the Truando to be five feet deep, and thirty-five yards wide, flowing between alluvial and level banks, elevated some eight feet above the river, and everywhere closely covered with vegetation. It appeared about one-third larger than the Nerqua, in the lower part of its course, and, like that river, its banks are a mere wilderness, inhabited by only one Indian and his family. It flows with a slow current, sixty feet in thirty-five seconds; and its bottom consists of rolled fragments of a clay slate, and associated deposits similar to those found in the channel of the Nerqua, with about the same proportion of pebbles of quartz, or indurated clay-stone streaked with quartz.

We left the confluence of the Nerqua and Truando at forty-five minutes past eight o'clock, A. M., and at twenty-five minutes past ten, after traversing some fine smooth reaches of the river, of a depth of five to six feet, bottom of shelving stratified rock, inclined at an acute angle, we arrived at the head of the first of the series of saltos, or rapids, which constitute the main difficulty and danger that we had to encounter in descending the Truando.

Five hours and twenty minutes were occupied by our people in bringing the baggage and the two canoes over the first four rapids. The operation was one of immense labor and difficulty: it became necessary or them to unload the canoes three times, and carry the cargoes by land, past the most dangerous places; and in bringing down the empty canoes, specially the large one, every effort of strength and skill was required in order to prevent them from being dashed to pieces against the rocks.

The valley through which this defile passes is very narrow, and hemmed in on either side by precipices of irregular unstratified rock.

At the foot of the fourth rapid we erected Rancho No. 7, on a rock close to the stream, and high enough to be beyond the reach of any sudden rise of the river. The barometer stood, at seven o'clock, P. M., 29 22-100, and the thermometer at 78 deg.

Between the second and third rapids the river receives a tributary on the right bank, of inconsiderable size, not exceeding at its junction four feet in width, and only a few inches deep. Aneroid stood, at half-past seven o'clock, at 29 25-100; the thermometer at 78 deg.

February 18.—At eight o'clock, A. M., we crossed the river in the large canoe, just above the fifth rapid, and made our way along the rocks which margin the river for the distance of half a mile, in which space there is a rectangular bend, and reached the main fall, which was twenty-six feet.

It rained very heavily, during which the people were employed at their difficult and hazardous occupation; they first emptied the canoes, and carried the things down by land to the foot of the rapid, and then brought down the empty canoes over rocks, by sliding them over pieces of wood, in which operation all our party cheerfully assisted. Below the main fall, we passed three other rapids in the same manner. About five o'clock P. M. the whole work was completed without the loss or damage of a single article.

I must here state that the manner in which this task was accomplished by Alexandro and his Indians elicited the warmest thanks, and the approbation of our whole party.

The barometer, at Rancho No. 8, at six o'clock, P. M., stood at 29 30-100. I ascertained the height of the main fall, by level, to be twenty-six feet four inches, and by aneroid, the same. At the head of the last rapid, the river receives a tributary from the left bank.

February 19.—It was my intention to have continued the descent of the river this day, but heavy rains had caused a rise of four or five feet; and as the channel was still narrow and rocky, Alexandro did not consider it prudent to venture, although another turn of the river would have brought us into perfectly smooth water.

February 20.—Last night the river very quickly subsided to its usual depth; but afterwards there was a heavy rain-storm, with thunder and lightning, and this morning it was higher than before, and much drift-wood floated down.

During our residence at this rancho, the aneroid barometer ranged from 29 25-100 to 29 40-100.

February 21.—The river has fallen, and at six o'clock A. M. we left Rancho No. 8, and dropped down the stream, and in a few minutes had cleared the narrow and rocky part of the channel, and reached the first of the wide and smooth reaches through which the Truando flows, after leaving the rapids. At twenty minutes past seven, I observed on the left bank the stratified clay slate, inclined north-east at an angle of 40 deg.; from this point the banks again became flat and alluvial.

At thirty-five minutes past seven, we passed the mouth of a river called Salida; opposite to its mouth is an island. The Salida enters the left side of the river, flowing south-east.

The Truando now passes through the flat alluvial plain of the Atrato. Its banks consist of alluvium entirely, and are, doubtless, very fertile. At seven o'clock, A. M., we passed the brazo, or arm, called Chuparador, which forms an island, and at twenty-five minutes past eleven we passed another island.

At half-past eleven we arrived at the head of the Pallisadas or series of islands, which owe their origin to the accumulation of drift-wood carried down by the floods. The branches of the river diverge, and inclose, like a net-work, the numerous wooded islands which compose this group, extending to a distance of four miles.

At the head of the Pallisadas, it was necessary to drag the canoes over a barrier of drift-wood several yards in breadth; but after this obstacle was overcome, we encountered no other in the descent of the river, except single trees crossing the channel, one or two of which we were compelled to cut through with the axe. At twenty-five minutes past three, we cut through the last tree, and then continued the descent without stoppage, until dark, when we halted for the night at a part where the river's banks were overflowed, which compelled us to pass the night in the canoes, in the best manner we could. The breadth of the river here is about twenty-five yards, with a forest of lofty trees on each side.

We passed about half a dozen uninhabited ranchos, built by Indians and others from the Atrato, who occasionally ascend the Truando on fishing and hunting excursions, as far as the Pallisadas, and even to the falls.

The distance run during the day, computing stoppages, flow of current, and rate of passage through the water, is shown on the accompanying table.

February 22.—We left our Rancho, No. 9, at ten minutes past six o'clock, and resumed our voyage down the Truando, passing a succession of slow and smooth reaches.

At fifty minutes past seven, A. M., we entered the first of the lagoons, and found a wide and open space, oblong in form, and ten miles in diameter. The whole of this open, circular space, is covered with the lagoon grass or sedge, which takes root below the surface of the water, and rises to the height of several feet above it. Through this green expanse flows the Truando, and other streams and brazos—the margins of the rivers being frequently rather above the level of the surrounding lagoon, supporting narrow belts of trees. A range of mountains is seen about twenty miles distant, to the north, and running north-east and south-west. At half-past nine we passed the first lagoon, and there the Truando re-entered a forest and assumed its previous form, no open space being visible on either side. At forty-five minutes past eleven, we passed the mouth of the Chuparador, about twenty yards wide, and flowing north-east and south-west. At half-past one we entered the second lagoon, which closely resembled the first one in its general aspect and configuration. The Truando, from this lagoon, resumes its former course, precisely. Owing to the swollen state of the Atrato, the Truando, during the latter part of this day, had no perceptible current; consequently, our canoes were paddled along leisurely. Finally, at five o'clock, P. M., we reached the Atrato.

The mouth of the Truando is about fifty yards in width, but owing to the peculiar appearance of the banks of the Atrato, it is scarcely visible from the latter river. The Truando is of the depth of twenty-seven feet inside of the bar; on the bar there are but eighteen feet. The width of the bar is thirty feet.

At forty-five minutes past five o'clock we reached the small pueblo, which is situated about two hundred yards below the mouth of the Sucio. Here the aneroid barometer stood at 29 37-100, at 6h. 45m., P. M. From the mouth of the Truando to that of the Sucio, is one mile and a-half.

From this point we proceeded to Quibdo, the capital of the Province of Choco, where we arrived on the tenth of March; and having remained there a month and four days, on business not connected with the researches of this expedition, we proceeded down the river Atrato, and arrived at its confluence with the Truando, at 6h. 30m., A. M., April 18th, 1855, from which point I began that part of my survey which remained unfinished, viz.: The river Atrato from the Truando confluence to its mouth. I sounded the river Atrato at its junction with the Truando, and found its depth to be fifty-eight feet; its width at this point is three hundred and fifty yards. The reach of the Atrato, at the Truando, runs due north.

At 7h. 10m., A. M., I arrived at the river Sucio. A sail was set on the vessel at 10h. 30m., A. M., which added to her speed. She went at the rate of eighty-eight feet in fourteen seconds, for thirty minutes. (See tables of Atrato.)

At 11h. 45m. we passed the river Salaqui, which river is called by the natives Leonda or Yegenta; it is shallow at its mouth, but opens into a large river. 7h. 50m., P. M., passed the mouth of the river Chercarita.

April 19.—4h. we passed the river Leonda, on the eastern bank. At 8h. A. M. we entered the first mouth of the river called Chocorita, and at 10h. 3m., A. M., we entered the Gulf of Darien, through the mouth of the Coquito.

After my arrival in New-York, on the 27th of May, 1855, I reduced my notes to mathematical calculations and drawings, and find, that not only a canal without locks is practicable, from the Pacific to the Atrato, but a navigable river can be turned, from the lagoons of the Atrato, into the Pacific, of sufficient magnitude to convey the commerce of the whole world.

My endeavors to point out the practicability of such an operation are expressed in the five Plates, with explanations of the same, which I have the honor to transmit herewith.

As regards the salubrity of the route, I have only to call your attention to my meteorological observations, and to the fact, that neither myself nor any of the party were sick, with the exception of a slight attack of fever and ague that I experienced while on the Nerqua, brought on, no doubt, from exertion and exposure; but from that time until my return to New-York, I enjoyed excellent health.

In respect to obtaining laborers, the four adjacent provinces can supply a great number of men to carry out the project, some of whom, in the province of Carthagena, have been employed on the Panama Railroad.

I would suggest that the undertaking be carried on in six divisions, stationed thus:—

The First Division at the mouth of the river Atrato.
The Second Division at the Truando.
The Third Division at Townsend Station.
The Fourth Division at the eastern mouth of the tunnel.

The Fifth Division at the western mouth of the tunnel, and
The Sixth Division at Kelley's Inlet.

To provide accommodation for such a number of men as will be necessary to construct this work, houses must be built. The facilities offered for this purpose is the palm-tree, which grows most abundantly in this part of the country. Ranchos built of palm are applicable to the Third, Fourth, Fifth and Sixth Divisions.

Those required for the Second Division (on the Truando), and for the First Division (on the Atrato), must be constructed to float, as there are no banks on those rivers on which ranchos, or houses, can be erected. The dredging machines to be used on the river can be made sufficiently capacious to accommodate all the men in each division.

East of the summit the best sites for the erection of work-shops and provision warehouses, are, Turbo, near the mouth of the Atrato, and Townsend Station, near the proposed junction of the canal with the Truando. Communication could be kept up between these two points, in twelve hours, by means of steamers drawing six feet of water.

West of the summit I would propose, as the best locations for erecting provision magazines, &c., the Gooding and Emma Josephine Stations to supply the Fourth and Fifth Divisions, and Kelley's Inlet to supply the Sixth Division. All the vegetables required could be raised along the shores of the Pacific, and in the valley of the Nerqua, after about a year's cultivation. The plantain, which is the chief food of the natives, grows here in great abundance; also the Indian corn and rice, to some extent, in the vicinity of the route, together with nearly all the fruits and esculent roots of the tropics.

Breadstuffs can be procured from Peru and Chili, as well as from San Francisco. Cattle, pork, &c., can be conveyed from Chiriqui to Kelley's Inlet in three or four days.

Thus a sufficient supply of victuals could be obtained on the Pacific Coast for the use of the Western Divisions.

Previous to the construction of a mule road across the summit, by which means fresh provisions, &c., could be carried over to the river Truando, from the same source as that at the mouth of the Atrato, namely: from the Provision Magazine at Turbo, which Depot would be supplied from New-York, *via* Carthagena.

As before stated, a steamer can make a trip to Townsend's Junction, from Turbo, in twenty-hours; and from Turbo to Carthagena in sixteen hours, even in the present condition of the country.

From my long experience, I am quite satisfied that colored labor is by far the most advantageous and preferable, as white men cannot perform the same amount of work in this country.

Before concluding, I have much pleasure in stating that my assistants, Mr. Norman Rude, Dr. R. G. Jameson and Mr. Robert Nelson, rendered me most valuable assistance during this arduous undertaking.

Having thus given you a full and detailed account of my journey, as well as my opinion of the practicability of the enterprise, and believing that it will sustain, and hoping that it may meet with, an impartial investigation, I have the honor to subscribe myself,

Your obedient servant,
WILLIAM KENNISH, *Chief Engineer*.

NEW-YORK, *August 7th*, 1855.

Panama, *25th April*, 1855.

WM. KENNISH, Esq.

Dear Sir:

As I have no doubt but that you will have arrived in the city of New-York by this time, and be astonished that you have received no letter from me, I am sorry to say that, unfortunately, I have been detained in Jurador much longer than I either wished or anticipated, whence I arrived here only yesterday, which will account for the delay.

According to the instructions which you gave me when we parted at the falls of Truando, after resting a few days at Paracuchichi, I proceeded to Jurador River, to examine it and the Indian path across the mountains to the Salaqui River, which flows into the Atrato.

February 28, 1855.—At Alexandro's house the barometer stood at 29.38, and the thermometer at 76 deg., at seven o'clock, A. M.; and the rest of the day the barometer at 29.45, and thermometer at 81 deg.

March 1.—Same place, at six o'clock, A. M., the barometer stood at 29.46, thermometer at 78 deg.; and at noon barometer stood at 29.50, and thermometer at 85 deg.

March 2.—At six o'clock, A. M., barometer stood at 29.42, and the thermometer at 74 deg. At seven, A. M., accompanied by a guide and two Indians, I started up the river in a small canoe, at the rate of about a mile an hour. Current runs at from two to three knots; course N. by W. At 3 o'clock, P. M., arrived at Dos Bocas (or the two mouths), so called from the junction of another river with the Jurador, of about the same size. Here we stopped for the night, having traveled about eight miles by river, or six in a straight line. Barometer 29.4; thermometer 82 deg. Six o'clock, P. M., barometer 29.41; thermometer 78 deg.

March 3.—Barometer 29.4; thermometer 72 deg. Started at seven, A. M., and traveled until three, P. M., when we arrived at a small river called Antado—general course N. by E.; distance twelve miles by river, or seven and one-half in a straight line. River much more winding, and only about half the size it was the day before—mountains on each side. Barometer 29.3; thermometer 82 deg. Six o'clock, P. M., barometer 29.25; thermometer 80 deg.

March 4.—Six, A. M., barometer 29.287; thermometer 75 deg. The river Antado is about the same size, and just such another river as the Hingador, near the Nerqua. The bed is full of stones and rocks, and as it is too small for canoes, we had to abandon them, and travel on foot, which we did for one mile 2,425 feet, general course east, until we came to a small brook, up which we went 2,000 feet, course northerly, when we arrived at the road to Salaqui. Barometer 29.22; thermometer 82 deg. We then ascended a high and pretty steep mountain, for about one mile, which is the summit. Barometer 28.5; thermometer 82 deg. Traveled along the ridge N. E., three miles or more, when we descended a quarter of a mile to a river called Mojando. Twelve o'clock M. barometer 28.95; thermometer 84 deg. From Mojando to the Salaqui River (4 1-2 miles N. E.) is a perfect level, with scarce an elevation—in some places swampy. At Salaqui barometer 28.95; thermometer 82 deg.

March 5.—Six o'clock, A. M., barometer 28.92 ; thermometer 73 deg. After the Indians had brought an old boat they had hidden somewhere in the river, we started down and arrived at the rapids next day about ten o'clock, A. M.—general course easterly ; distance, in a straight line, about twelve miles. The Salaqui River, in everything, has the same appearance as the Truando. It has the same depth, and runs through a beautiful valley like the Nerqua. Barometer 29.06 ; thermometer 85 deg.

As the canoe was too large to drag down the rapids, we left it at the top, and walked or crept down the side of the river for about three miles more, but did not see any fall, such as the great falls on the Truando, and have been informed that there are none but rapids.

Having come to a part of the river that we could not pass without a canoe, we were obliged to return. Barometer 29.1 ; thermometer 82 deg.

The mountains on each side of the rapids have the same appearance, and are of about the same heights as those on the Truando ; but as they recede at an angle of about 25 to 40 deg., I think they could be cut down so as to make sufficient room for both canal and river.

March 11.—We again arrived at Alexandro's house.

Thus I have given you an account of my journey to Salaqui, with observations, and leave you to form your own opinion, having no doubt but you will agree with me in preferring the Nerqua road, being by much the shortest and easiest.

I inquired at Santiago about the short road that Alexandro said he knew, but he protested that there was no such road.

(Signed,)

ROBERT NELSON.

THE RIVERS JURADOR AND SALAQUI.

Courses and Distances.

Names of Places.	Course.	Distance. Miles.	Feet.
From Alexandro's house to Dos Bocas...............	N. by W.	8	....
" Dos Bocas to river Antado.....................	N. by E.	12	....
" River Antado to Summit......................	N. E.	1	2,425
" " "	N.	–	2,000
Ascent of Summit.................................	..	1	....
From Summit to Mojando..........................	N. E.	3	1,320
" Mojando to Salaqui...........................	E. N. E.	4	2,640
		30	3,105
Descent of Salaqui to Rapids.......................	E.	12	....
" " below Rapids.....................	E.	3	....
		15	....

THE RIVERS JURADOR AND SALAQUI.

Barometrical Observations.

Place of Observation.	Day and Hour.			Bar.	Ther.
Alexandro's House,............	Feb'y 28th,	7 A. M.,		29.380	76°
" "	Rest of day,			29.450	81°
" "	March 1st,	6 A. M.,		29.460	78°
" "	" "	12 M.,		29.500	85°
" "	" 2d,	6 A. M.,		29.425	74°
Dos Bocas,....................	" 3d,	7 A. M.,		29.400	72°
River Antado,.................	" "	3 P. M.,		29.300	82°
" "	" "	6 P. M.,		29.250	80°
" "	" 4th,	6 A. M.,		29.287	75°
Road to Salaqui,..............	" "	8 A. M.,		29.225	82°
Summit,.......................	" "	10 A. M.,		28.500	82°
Mojando,.....................	" "	12 M.,		29.950	84°
Salaqui,......................	" "	3 P. M.,		28.950	82°
"	" 5th,	6 A. M.,		28.925	73°
"	" "	10 A. M.,		29.060	85°
"	" "	4 P. M.,		29.100	82°

(Signed,)

ROBERT NELSON.

THE INTER-OCEANIC AQUEDUCT.

WM. KENNISH, *Chief Engineer.*

EXPLANATION OF PLATES.

Plate I.—This Plate shows a section of the Inter-Oceanic River, from Kelley's Inlet on the Pacific, to the junction of the rivers Truando and Atrato. The distance and elevation of the tidal influence on the Mary River was measured, and the line of level sprung from that point, allowing for the fall of the tide (the level being taken at high water), from thence to Rancho No. 1, and on to Dos Bocas, without interruption, as shown on the Section and Plate 3 ; but meeting at the last-named point with a densely wooded valley—through which to cut a path to afford an opportunity to get a clear sight, would have consumed more time than could then be afforded—it was thought advisable to run a line of levels along a ridge of oblong hills (shown in Plate 3), nearly north-east and south-west, leaving a valley on either side—the valley on the right being the deepest. On this ridge of hills, represented by a dotted line from Dos Bocas, over the summit, is shown the height of leveling at each number; and the true profile of the proposed route was determined by offsets transverse to the general direction; the angles being taken, and a measurement by tape, from each reduced level to the bottom of the valley, as shown by the perpendicular dotted lines—the distance being calculated in the usual manner. At the Chupipi the level agreed with the barometer. (See Barometrical Observations.)

From the summit to the river Nerqua is an abrupt descent; the angle of depression was taken, and the distance from the summit to the Nerqua measured accordingly.

At Rancho No. 5 a series of barometrical observations were taken, and compared with the Chupipi, or Rancho No. 2, to endeavor to check the elevation; which two points were found to be on a level within a few feet. The distance and descent from Rancho No. 5 to the junction of the Truando with the Atrato, was ascertained from the different rates of current. (See Tables.)

The red coloring, on Plate No. 1, shows the cutting to be made from the Pacific to the junction of the New River and Truando; the same color, shown at the bottom of the Truando, represents the dredging required to deepen that river to an uniform depth of thirty feet.

Of the two lines represented on Plate 1, as running over the Truando, the black line shows the present surface of the river, and the red line the proposed surface, when deepened: on the bottom of the Truando, the black irregular line shows the present depth, and the straight red line the proposed depth.

The blue line passing from the Pacific to the Atrato, is the datum, or mean level line, of the two oceans.

The height of the tide on the Pacific at that point is 12 ft. 6 in., that is to say, 6 ft. 3 in. above, and 6 ft. 3 in. below the mean level; which latter would be the 36 ft. 3 in. above the bottom of the canal at this point.

The fall from the summit level to the Pacific would be, at low water, 15.2 feet plus 6.25 feet, and at high water, 15.2 feet minus 6.25 feet.

The red perpendicular lines show the depth of cutting, divided into sections, and the black figures standing on an angle from the datum-line show the distances from the Pacific to each point indicated.

It was thought advisable to recommend two tunnels in lieu of one, for the following reasons: the width of the tunnel must be equal to that of the river (200 feet), to prevent an increase in the current, and one arch sprung over this immense width would be too expansive to ensure safety; therefore two arches were substituted.

This division will also prove advantageous to the transit, as it would preclude all danger of collision, by the ships observing the present law of navigation, viz.: that of keeping to the right.

As to the heights of the proposed tunnels, they are quite sufficient to allow a line-of-battle-ship to pass through, with her top-mast and top-gallant mast struck, and her yards braced to within the limit of the width of the vessel.

Sailing frigates, and merchantmen generally, need only lower their top-mast, or slack it away a little, in the mast ropes.

All steamers can pass through without altering any of their gear.

By referring to the Table of Calculation for ascertaining the solid contents of matter to be removed in cutting the bed of the river, it will be seen that almost two-thirds is rock, which will form a permanent wall on either side, and preclude the necessity of forming an embankment of masonry, and at the same time save an immense quantity of cutting, as the sides of the river-bed need not exceed an angle of 30 deg. from the perpendicular, and the rock being of a primitive formation, or trap, it will yield easily to the blast: at the same time rock is easier to be removed to the adjacent valleys than a softer material.

There is sufficient water in this vicinity to erect hydraulic works, for hoisting rock and other material over the bank.

A height of from five hundred to six hundred feet of water can be obtained at each end of the tunnel; this, led through pipes to the canal, would work machines sufficient to perform any required operations.

WILLIAM KENNISH,
Chief Engineer.

AUGUST 7, 1855.

A DESCRIPTION

Of the proposed New River Aqueduct, which is intended to flow from the lagoons of the river Atrato into the Pacific Ocean, thus forming an uninterrupted communication with the Atlantic Ocean, through which the largest ships may pass, and with a current of not more than two miles per hour.

The principles on which this desideratum is to be effected, are:—

First.—That the mean level tide of both oceans is the same.

Second.—That the tidal wave at the mouth of the New River, on the Pacific shore, oscillates twelve feet six inches at spring tides, and ten feet eleven inches at neap tides; while at the mouth of the river Atrato (the terminus of the passage on the Atlantic), the tidal wave does not exceed two feet at any phase of the moon.

Third.—That the waters of the Atrato, at the points of junction with the New River, are fifteen feet, two-tenths, above the mean tidal level of either ocean.

Fourth.—That the junction or summit of the New River, is, therefore, nine feet above the Pacific at the highest tide, and thus the water will flow down it with a velocity equal to that head; while at the lowest tide the velocity will be equal to twenty-one and forty-five hundredths feet head. (See Explanation of Section.)

The length of the river Atrato from the proposed junction to its mouth, is sixty-one miles six hundred and sixty five yards, and that of the New River, sixty-three miles ten hundred and sixteen yards; their rate of current will, consequently, be nearly the same. The time and power, therefore, expended in ascending the Atrato will be compensated in descending the New River, and contra-wise; making it equal to sailing a distance of one hundred and twenty-five miles on a level surface.

The summit formed by the Atrato and the head of the New River, being of the same height from the mean level of either ocean, the tidal influence of the Pacific will ascend the New River, nearly to the summit at every high water, and cause an irregularity in the current, whose average rate will be nearly equal to that of the Atrato, (two and a-half miles per hour.)

As this river is intended to flow continually into the Pacific, without obstruction, so that vessels may at all times pass through, it may be necessary to state whence the source from which a sufficient supply of water is to be obtained.

In the first place, it may be proper to point out the geographical position of the country.

The Cordilleras of South America run parallel with the Pacific coast in the vicinity of the head waters of the Atrato, in a N. W. direction, toward the mouth of that river; and from these, springs another range of high hills, called the Antiochian Mountains, running in a N. E. direction. Between these two ranges of lofty mountains lies the vast alluvial valley of the Atrato, which river, during the lapse of ages, has formed for itself a deep and spacious bed, running about midway between the hills, and for a distance of three hundred miles.

The width of the Atrato Valley, opposite the proposed junction, is about fifty miles.

This vast area may be said to be covered with water, with the exception of here and there a few islands, thrown up by the currents of the various smaller streams in their winding course to the great artery.

The rivers flowing from the Antiochian Mountains invariably preserve a north-west, and those from the Cordilleras a north-east course, pouring all their waters into the Atrato River, which, in its turn, overflowing its banks, inundates the surrounding country, forming deep and widespread lagoons on either side.

It is known as an established fact, that the clouds seldom pass over the Cordilleras, toward the Pacific, but are attracted by the mountains, and disgorge themselves on the Atlantic side; hence the reason of the perpetual rain, thunder, and lightning in the Atrato Valley, while on the Pacific coast there is scarcely any rain for eight months in the year—thus accounting for the absence of rivers on that coast, in this section of the country. (See Plate 6.)

From the fact that the greater portion of this rain falls above the proposed junction, and is carried into the Atrato by upwards of fifteen large tributaries, besides numerous small streams, and also from the fact that these lagoons, or immense lakes, covering hundreds of acres, and maintaining an undiminished supply through all seasons of the year, are likewise above this level, it is sufficiently obvious that a superabundance of water may be constantly depended on.

This most important point being established, it only remains to prove that the place selected on the Atrato is the most practicable for the purpose.

In the first place, there is no point of junction with this river, by others flowing from the west, so near the level of high water on the Pacific, as that of the Truando, which is just three feet above that line, and thus of sufficient height to prevent the Pacific, at high water, from flowing to the Atrato, and not so high as to cause a too rapid current through the New River at low water; in fact, just preserving a preponderating balance on the side of the former.

Should either of the rivers Pocador or Napipi be selected, the former being twenty-eight feet above the mean tidal level, and the latter thirty feet, and the rise and fall of the tide in Capica Bay being sixteen feet, the Pocador would, consequently, be thirty-six feet above low-water level, and twenty feet above high water, and the Napipi thirty-eight feet above low-water level, and twenty-two feet above high water. A canal therefore made in the vicinity of these rivers, would have a tidal current of about five miles per hour.

The Atrato, at the junction of the Salaqui, is only one foot above the level of high water on the Pacific; but the dividing ridge is one thousand and sixty-three feet high, and thirty miles in width, from the Jurador to the opposite side of the Cordilleras. (See Plate.) This line was partly run by myself, and afterwards finished by Mr. Robert Nelson, my assistant. (See Nelson's letter, annexed.)

Again; should any of the rivers at the mouth of the Atrato be selected, without reference to the height and width of the dividing ridge, I may observe, that the tidal wave on the Pacific is twenty-three feet at high water above the Atlantic, opposite the mouth of the Atrato, in a southwest direction, which would be in the vicinity of the river Sugra, or its tributaries, Maria, Balsa, Yarisa, &c., all emptying into the Gulf of San Miguel. The tide also rises from twenty-five to thirty feet at Chipagana, a small town situate eight or nine miles above the mouth of the river Savanna; while at the mouth of the Atrato, as before observed, the rise and fall are but two feet. Taking the maximum of the tidal wave on the Pacific, then, to be twenty-five feet, and that on the Atlantic two feet, the tide at high water would flow from the Pacific to the Atrato, with a current equal to a head of eleven and one-half feet, and at low water on the Pacific the water would flow from the Atlantic with a current also equal to eleven and a-half feet. In other words, the Pacific would flow into the Atlantic for six hours, and the Atlantic into the Pacific for six hours.

Notwithstanding the dividing ridge has never been explored, I have heard extraordinary stories from the natives concerning that region of the country. During my three visits to the Gulf of San Miguel and vicinity, I was told the San Blas Indians communicated with those of the Arquian territory, and that in an extraordinarily short time. But my opinion of this matter is, that all information derived from the natives is to be received with great caution: the truth or falsity of their assertions can only be proved by actual survey.

Here I would mention, for general information, that, in the so (lately) called Darien harbor, (an arm of the sea, passing through Boca Grande and Boca Chica,) two small inlets from the Gulf of San Miguel, passing up the river Savanna for about twenty miles, and the Sugra as far as Yarisa, some twenty or thirty miles, that the action of the tide is such on ships at anchor, that while I was on board of H. B. M.'s steamship Virago, Capt. Prevost, the first and only ship that ever entered that inlet, she dragged both anchors ahead, and was only brought up by paying out nearly all the cable she had. There is scarcely a place within the two inlets above mentioned where a ship can anchor with safety, owing to the action of the tide.

WILLIAM KENNISH,
Chief Engineer.

AUGUST 7, 1855.

THEORY

Of the United Action of the Tide on the Pacific and the Summit Level of the New River, at the Lagoons of the Atrato.

This is a scientific problem that alone involves practical knowledge, inasmuch as there is no formula given by any writer on hydraulics, extant, to elucidate the required results.

Therefore this most important subject must be treated from analogical inferences alone.

The points to be demonstrated are these:—Will the fresh water from the summit continue to flow toward the Pacific, at the top of high water, with a velocity due to a head of nine feet—the height of that summit being nine feet above the high-water line on that coast? Although the tide on the Pacific has risen twelve feet six inches from the low-water mark, will the current in the river run with the velocity that a head of twenty-one feet would give it at low water? Or, will the salt water from the ocean, by the influx of the tide, maintain the lower regions of the river, in consequence of its superior gravity, and lift the fresh water, without materially commingling, (being of an inferior specific gravity,) to the height of the tides, without interfering with the current of the river more than the difference of the heads presented at the summit to the different stages of the flux and reflux of the tide on the Pacific?

To answer the above queries, reference must be had to such natural rivers as are similarly situated with regard to the tide.

London, on the Thames, is situated about fifty miles from the Nore, or the mouth of that river. The tidal wave reaches London Bridge three hours and thirty minutes after passing Dover harbor, about seventy-two miles from London; from which fact it is evident that the tide, at the mouth of the river, is half ebb when it is high water at London Bridge, and that, when it is low water at London Bridge, it is half flood at the Nore, or mouth of the river—thus preventing that current that would otherwise exist in the river at low water, should the tide begin to fall at London Bridge at the same time as it does at the Nore. But the tide in the Thames is acted on by two contrary ocean tides, viz.: that flowing southerly, along the eastern shores of Scotland, and that flowing northerly, through the Straits of Dover, which, meeting, send the flood up the Thames with great force. All tidal rivers more or less partake of these qualifications, according to the features of their adjacent shores; but, in the case of the proposed New River, it is materially altered, inasmuch as its bed is to be uniform in width and depth, from end to end; therefore, the current from the summit to the Pacific will (unobstructed by tortuous bends, which invariably exist in natural rivers) preserve a uniform flow, on known hydraulic principles, and at the same time admit the passage of the tide under the fresh water coming down the river, without antagonistic commotion either at the flood or ebb. The river will flow with a smooth current of about two miles per hour, and which, as far as theory can elucidate the tidal influence, will extend to the Townsend Junction nearly. From thence to the summit the water will flow toward the Pacific nearly at the same rate of current as the Atrato to the Atlantic. That part of the river between Townsend Junction and the Pacific will be slightly agitated by the rise and fall of the tide, but the velocity of the current will be scarcely affected.

As this portion of the country has never before been visited by a white man of any nation, my route is here, doubtless, open to more minute exploration, and a more careful survey may suggest Humboldt's Bay as the Pacific terminus; in which case the only obstacles to be encountered in connecting the river with that, or the Corredor, are the expense of excavation from thence to Kelley's Inlet, and the crossing of the river Paracuchichi. As far, however, as hydraulic principles are concerned, I believe what I offer to be as true as instruments and scientific formulæ can effect it.

As this is the grand basis upon which I have endeavored to raise my structure for accomplishing the object in view, viz.:—to join the two great oceans by a never failing river, any small discrepancies in my given heights or distances, are obviously unimportant; for the difference of a few feet, more or less, in my elevation of the summit, of either ocean, could easily be met by a slight modification of the line, and would in no way prejudice the plan of connecting the two great waters by a canal without locks.

In my proofs of the practicability of this I allow of no discrepancy, the data being founded on that immutable law of nature, *gravitation*.

WILLIAM KENNISH, *Chief Engineer*.

NEW-YORK, August 7, 1855.

[Omitted on page 11, after the fifth paragraph.]

The winds on the Pacific coast are periodical, viz., blowing from the south from March to September, and from the north from September to March; and during the whole year it is calm from 6, P. M., to 10, A. M., a gentle gale beginning at 10, A. M., which increases to a topgallant-sail breeze until noon, when it gradually dies away to a calm at 6, P. M.

Consequently, a vessel, once towed to a fair offing, can depend on a breeze of four knots per hour for six hours during the twenty-four, which will bring the vessel into the general trade winds, clear of all head winds, in a few days.

ERRATA.

On page 6, fifth line from bottom, for "hatchets," read *machetes*.
On page 8, on twenty-fifth line from top, for "it is possible," read, it is *not* possible.
On page 9, on tenth line from top, for "Leewardly," read *Seawardly*.

HEIGHT OF THE ATRATO AT ITS JUNCTION WITH THE TRUANDO.

The mean central current of the river is about 2.7 miles per hour, with a mean hydraulic depth of 53 feet, and the formula upon which the calculation is based is, that the square root of the mean hydraulic depth, multiplied by the fall—both in feet—will give the mean velocity of the current, in miles, per hour, under a deduction, according to the volume of the stream, the directness of its course, and the nature of its banks. This course of calculation has been thought preferable to that of adopting any of the more learned formulæ, as there are now so many well ascertained facts relative to most of the chief rivers of the world, that the proper deduction can be ascertained with much precision.

In Beardmore's tables of the discharge by water-courses of large area, the deduction is from 1-9th to 1-8th.

In the case of the Nile, the facts are not entirely reliable, but, as far as they go, would give a deduction at low water of 1-3d, and in flood of 1-6th.

The Neva, at St. Petersburgh, has a mean hydraulic depth, at a low state, of about 60 feet, and is, in many respects, similar to the Atrato. The facts relative to it have been well established, and the greatest deduction which the observations give is 2-5ths.

In the Ganges the deduction would be nearly 1-3d.

In the English rivers the deduction is rarely above 1-5th: and that only at a low state, *i. e.*, the greatest deduction. And this refers to nearly all the large streams.

In applying the formula to the Atrato, and taking the observed current as the mean velocity of the water, the rise would be only 9 1-2 feet: the correction in this case becomes an addition, as the current is one of the data. After much calculation, and a careful consideration of many observations of the Atrato, I have fixed the height at 15.2 feet. To apply the correction from the Neva, the height would be only 13 1-3d feet, and the corrections from any of the other rivers would make the height still less.

WILLIAM KENNISH, *Chief Engineer.*

CALCULATION

Of the discharge of the River Atrato, and the quantity of water required to supply the proposed

RIVER AQUEDUCT.

RIVER ATRATO.								
Cross Section at Truando.			Cross Section at 99 miles 1,546 yds. on Longl. Section.			Section at Cano Barbacoa.		
Lengths, Feet.	Depths, Feet.	Areas, Sq. Feet.	Lengths, Feet.	Depths, Feet.	Areas, Sq. Feet.	Lengths, Feet.	Depths, Feet.	Areas, Sq. Feet.
70	17	2,065	61	29	2,928	74	14	1,554
75	42	3,468	73	67	5,584	194	28	6,838
128	50½	6,656	124	86	10,974	108	42½	4,536
150	53½	8,137	123	91	11,808	119	41½	5,295
110	55	6,215	223	1,101	22,634	96	47½	4,584
168	58	9,408	94	102	9,306	224	48	10,192
211	54	10,233	97	96	8,584	196	43	7,840
24	43	2,867	80	81	5,400	136	37	4,148
64	34½	1,424	60	54	2,490	28	24	532
	10			29			14	
1,000		50,473	935		79,708	1,175		45,519

SUMMARY.

GREATEST DEPTHS.	AREAS.
Feet.	Sq. Feet.
58	50,473
102	79,708
48	45,519
208	175,700

The mean depth of the Atrato, from the Truando to its delta, being 78 feet:—if 208 : 175,700 :: 78=65,800 feet area, mean.

SECTION OF THE ATRATO.

The mean average surface current of the Atrato is 2.6 miles per hour, and the mean sectional velocity will therefore be 1.92 miles per hour, or 10,137 feet, which gives a discharge per hour of 667,014,600 cubic feet.

RIVER AQUEDUCT.

AT LOW WATER.

Fall, . . . 21.45 feet, } Fall, per mile, 0.382 feet.
Length, . 56.08 miles,

Sectional area, 5,000 sq. feet; mean hydraulic depth, 22.22 feet: hence surface velocity would be 2.549 miles per hour, and mean velocity 1.9 miles per hour, or 10,032 feet, giving a discharge of 50,160,000 feet per hour.

AT HIGH WATER.

Fall, . . . 9 feet, } Fall, per mile, 0.16 feet.
Length, . . 56.08 miles,

During this state of the river aqueduct, the lower portions would be increased in sectional area; but as the upper part would remain nearly the same as at low water, the section there would give the ruling depth for discharge.

The surface velocity would be equal to 1.7 miles per hour, and mean velocity, 6,864 feet, giving a discharge of 34,320,000 cubic feet per hour.

The mean discharge of the river aqueduct, per hour, would be about 42,000,000 of cubic feet; and supposing the whole of this quantity had to be got from the bed of the Atrato, it would reduce the discharge of that river one-sixteenth, and the level of the surface of the water about three feet and three-quarters. But, in fact, the greater portion of the water required would be got from the lagoons, which at present chiefly ooze out in numerous channels into the deltas of the Atrato, and partly pass off by evaporation.

The volume of water in the river aqueduct at high water, above that at low water, will be about 150,000,000 cubic feet; while the discharge, during the rise of the tide, would be about 250,000,000 of feet.

CUTTINGS IN BED OF TRUANDO.

Section K.....	Mean of extreme sections taken for Soil Cuttings, and greatest section of Rock by 1-3 the length for Rock Cutting.				
	Greatest Cross Section. Sq. Yds.	Least Cross Section. Sq. Yds.	Difference of Sections. Sq. Yds.	Difference of Centre Depths of Sections. Feet.	Yards, per foot of difference. Sq. Yds.
Section L......	860	745	115	16.5	7.
" M.....	743	619	124	17.0	7.3
" N......	618	474	144	14.5	10.
" O......	419	269	150	13.5	11.1

TABLE OF DISTANCES,

From Zero, on Pacific Ocean, along line of proposed Aqueduct.

	Distances between Stations.	Distances from Zero.
	Miles. Yds.	Miles. Yds.
From Zero to tidal influence,	2 0	2 0
Tidal influence to Rancho No. 1,	1 440	3 440
Rancho No. 1 to Dos Bocas,	0 1000	3 1440
Dos Bocas to Rancho No. 2,	1 970	5 650
Rancho No. 2 to Rancho No. 3,	2 888	7 1538
" No. 3 to Chuperador,	0 1257	8 1035
Chuperador to its Branch,	0 1360	9 635
Branch of Chuperador to Rancho No. 4,	1 115	10 750
Rancho No. 4 to Rancho No. 5,	1 880	11 1630
	11 1630	

NOTE.—The above distances, taken by actual measurement, correspond with those given in my daily journal, with the exception of that from Rancho No. 4 to No. 5, the distance having ceased to be taken in my line of survey, as stated in the above journal. After arriving at No. 5, the distance was measured backward to Rancho No. 4.

LINE OF LEVELINGS,

From Zero to the Summit.

Distances from Zero.	Heights of leveled points along ridge above datum-line.	Measured depths from leveled pts. to line of Section.	Heights in line of Section above datum-line.	STATIONS.	
Miles. Yds.	Feet. dec.	Feet. dec.	Feet. dec.		
2 0	11.4		11.4	Tidal influence extends.	
3 440	22.2		22.2	No. 1 Rancho and No. 1 of points on
3 1440	48.0	20.0	28.0	Dos Bocas,	No. 2. [ridge,
4 780	103.4	53.0	50.4		No. 3.
4 1300	128.5	55.0	73.5		No. 4.
5 650	172.8	79.0	93.8	No. 2 Rancho,	No. 5.
6 160	232.5	111.0	121.5		No. 6.
6 1520	301.0	150.0	151.0		No. 7.
7 680	443.0	284.0	159.0		No. 8.
7 1538	386.0	222.0	164.0	No. 3 Rancho,	No. 9.
8 1035	458.0	58.0	400.0	Rio Chuperador,	No. 10.
9 635	528.0	75.0	453.0	Branch do.	No. 11.
10 750	540.0	40.0	500.0	No. 4 Rancho and summit point of section.	No. 12,

RIO NERQUA,

From Rancho No. 5 to No. 6.

Stoppages, Rate of Current, &c., &c.

TIME.	STOPPAGES.	RATE of CURRENT.		DEPTH.	WIDTH.
Hours. Min.	Minutes.	Feet in seconds.		Feet.	Yards.
9.15	5	20	15	4	20
9.30	2½				
9.52	7½	20	45	5	18
10.2½	5	20	20		
10.10	3			7	17
10.20	8				
10.46	4	20	36	6	16
11.2	16				
11.10	2			8	17
11.33	5				
11.35	2			8	19
11.55	2	6	20		
12.13	4			9	20
1.23	25				
1.55	17			9	21
2.30	23				
3.25	5				
4.0	48	6	16	9	22

Total time,..............405 minutes.
" Stoppages,..........184 "
Remaining time,..........221 "
Mean current,...........75 feet per minute, gives whole distance 3 miles and 245 yards.

RIVER NERQUA,

From Rancho No. 6 to confluence with Truando.

TIME.	STOPPAGE.	RATE of CURRENT.	DEPTH.	WIDTH.
Hours. min.	Hours. min.	Feet in seconds.	Feet.	Yards.
6.45		60 16		
7.10	5	60 16	9	22
7.39	2	60 16		

Mean current,..............225 feet per minute.
Time,.................. 88 minutes.
Stoppages,............. 7   "
81 × 225 ft. = 3 miles 795 yds.

RIVER TRUANDO.

From confluence to first Salto.

TIME.	STOPPAGE.	RATE of CURRENT.	DEPTH.	WIDTH.
Hours. min.	Hours. Min.	Feet in seconds.	Feet.	Yards.
8.45		60 35	10	23
9.25		60 35		
10.25		60 35	12	30

Mean current,..............100 feet per minute.
Time,................100 minutes.
1 mile, 1573 yards.

The distance from first Salto to Rancho No. 7 was ascertained by actual measurement to be two miles. From Rancho No. 7 to No. 8 was 2½ miles. Whole distance from first Salto to Rancho No. 8, 4½ miles.

RIVER TRUANDO,

From Rancho No. 8 to No. 9.

Time, Rate of Current, &c.

Time.	Stoppage.	Depth.	Width.	Mean Rate of Current.		Mean Rate of Boat through Water.	
Hours. Min.	Hours. Min.	Feet.	Yds.	Feet in Sec.		Feet in Sec.	
6.27		9	30				
6.58	2						
7.17		10	33	28	35	28	35
7.35	32						
8.00		12	35				
8.5							
9.00		18	40	35	20		
9.15							
9.25		15	37	34	15	34	20
10.00							
10.6		10	40	28	15	28	15
10.7							
10.27	8	18	45				
10.52	19						
11.15		15	44	28	15	18	15
12.25	55						
12.40							
12.52		24	39				
1.15	2						
1.30	5			22	10	18	15
1.50	20						
2.10	6						
2.33		10					
2.50	16	15					
3.10	2	18					
3.15	5		40				
3.25	25						
3.48		26					
4.10		15	41	22	10	18	15
5.00		18					
5.16		21					
5.45		24	40				
6.00		18					
6.20		16					
6.24		20	39				
7.00		23					

Time,................... 753 minutes.
Stoppages,................ 197 "
Remaining time,........... 556 "
Mean current between Rancho Nos. 8 and 9, 95 feet per minute.
Mean rate through water, 75 feet per minute.
Whole distance from Rancho No. 8 to No. 9, 17 miles 1,586 yards.

RIVER TRUANDO,

From Rancho No. 9 to Atrato.

Time, Rate of Current, &c.

Time.	Stoppage.	Depth.	Width.	Mean Rate of Current.		Rate of Boat through Water.	
Hours. Min.	Hours. Min.	Feet.	Yds.	Feet in Sec.		Feet in Sec.	
6.10		18	41			24	15
7.00		16		30	10	18	15
7.40		22	40	26	15	22	15
7.45		14		20	15	18	15
8.40		15	38	20	15	16	15
9.15		16		30	15	26	15
10.40		20		32	15	16	15
11.45		18	40	22	15	16	15
12.15		21		20	15	22	15
1.15		20	39	17	15	24	15
2.00		26		15	15	26	15
2.12		23	39	16	15	22	15
3.14	.15	21		7	15	38	15
4.00		26	39	Nil.		38	15
5.00		27	40	"			

Time,.......... 10 h. 50 m.
Stoppages,..... 15 m. = 10 h. 35 m. × 172½ feet = 27 m.160 yds.
Mean rate of current, 172½ feet per minute.
   "   " through water, 172½ feet per minute.

RIVER ATRATO.

Time, Rate of Current, Calculated Distances, &c.

	Time.	Mean Current.	Depth.	Calculated Distances.		Distances from Truando.		Distances from Zero on Pacific.	
	Hrs. Min.	Ft. per min.	In Feet.	Miles.	Yds.	Miles.	Yds.	Miles.	Yds.
Stoppage,	6.30		58					63	1216
0 h. 45 m.	7.10	207	78	1	1011	1	1011	65	467
	8.43	207	120	1	1566	3	817	67	273
	9.20	253	66	1	1358	4	415	68	1631
Width,	10.30	245	54	3	620	8	1035	72	491
340 yds.	11.00	212	78	2	251	10	1286	74	742
	1.00	233	78	4	1445	15	971	79	427
	3.15	207	84	5	796	20	1717	84	1173
	4.00	232	84	1	1743	22	1700	86	1156
	4.30	306	66	1	318	24	258	87	1474
	4.45	285	68	0	1232	24	1490	88	946
Width,	5.00	212	66	0	1532	25	1266	89	718
400 yds.	6.00	201	75	3	420	28	1682	92	1138
	9.00	169	102	7	408	36	330	99	1546
	12.00	197	90	6	1525	43	95	106	1311
	2.00	197	78	3	1491	46	1586	110	1042
	7.00	197	69	11	354	58	180	121	1396
Width,	7.30	197	54	1	1068	59	1248	123	704
350 yds.	8.00	197	48	1	1068	61	556	125	12
	To delta of Atrato,......			61	556	61	556	125	12
	Cano Barbacoas to mouth of Coquito,..........			4	880	65	1436	129	892
	Cano Coquito to Gulf of Darien,.............			2	000	67	1436	131	892

TABLE OF BAROMETRICAL OBSERVATIONS.

42

PLACE.	DAY AND HOUR.	Bar.	Ther.	DAY AND HOUR.	Bar.	Ther.	REMARKS.
Paracuchichi,	Jan. 15, 6 P. M.	29.350	78°	Jan. 16, 6 A. M.	29.300	78°	
"	" 17, 6 A. M.	29.350	74°	" 17, 6 P. M.	29.300	80°	
"	" 18, 6 "	29.350	75°	" 18, 6 "	29.300	78°	
"	" 19, 6 "	29.350	75°	" 19, 6 "	29.300	78°	
"	" 20, 6 "	29.300	75°	" 20, 6 "	29.300	78°	
"	" 21, 10 "	29.350	78°	" 21, 6 "	29.300	78°	
"	" 22, 6 "	29.350	76°	" 22, 10 A. M.	29.400	76°	
"	" 23, 6 "	29.400	76°	" 23, 8 "	29.450	80°	
"	" 24, 6 "	29.400	75°	" 24, 8 "	29.450	84°	
"	" 25, 6 "	29.350	74°	" 25, 8 "	29.370	77°	
"	" 26, 6 "	29.350	73°	" 26, 8 "	29.400	80°	
"	" 27, 6 "	29.35	73°	" 27, 8 "	29.400	80°	
"	" 31, 1 P. M.	29.410	80°	" 31, 6 P. M.	29.35		
Rancho No. 1,	" 31, 4 "	29.385		" 31, 4 "	29.385		
Chupipi R. No. 2,	" 31, 4 A. M.	29.115		" 31, 6 A. M.	29.150		
Rancho No. 3,	Feb. 1, 8 A. M.	28.600		Chupipi, 6 "	29.13	76°	
" No. 4,	" 1, 10 "	28.650					Heavy rains every afternoon, attended with thunder and lightning.
Hing Falls, top,	" 8, 11 "	28.680					
" bottom,	" 8, 11.15 "	28.830					
Nerqua,	" 1, 6 "	29.225	80°	Feb. 1, 10 A. M.	29.300	80°	
Ranch No. 5,	" 3, 6 "	29.225	75°	" 3, 4 P. M.	29.200	78°	
" No. 5,	" 4, 6 "	29.200	76°	" 4, 4 "	29.200	78°	
" No. 5,	" 5, 6 "	29.150	76°	" 5, 4 "	29.200	77°	
" No. 5,	" 8, 3 P. M.	29.250	80°	" 9, 2 "	29.220	78°	
Nerqua R. No. 5,	" 10, 6 A. M.	29.200	74°	" 10, 10 A. M.	29.154	74°	During these days the mornings were cloudy, and no rain fell day or night; apparently the Verana was setting in. The observations of
" No. 5,	" 11, 6 "	29.200	72°	" 11, noon.	29.172	82°	
" No. 5,	" 11, 8½ P.M.	29.170	75°	" 11, 8½ A. M.	29.250	82°	
" No. 5,	" 12, 5½ "	29.175	70°	" 12, 6½ "	29.200	72°	
" No. 5,	" 12, 6 "	29.175	72°	" 12, 6 "	29.175	71°	

PLACE.	DAY AND HOUR.	BAR.	THER.	DAY AND HOUR.	BAR.	THER.	REMARKS.
Nerqua R., No. 5,.....	FEB. 13, 10 P. M.	29.200	Not noticed.	FEB. 14, 6 A. M.	29.150	71°	Mason's Hygrometer, taken nearly every day, indicated a difference between the two thermometers of from 1° to 5°, and in one instance of 7°.
" No. 5,.....	" 14, 6 A. M.	29.150	71°	" 14, noon.	29.118	85°	
" No. 5,.....	" 15, 6 "	29.125	70°	" 15, 10½ A. M.	29.175	80°	
" No. 5,.....	" 15, noon.	29.150	82°	" 15, 9 P. M.	29.200	Not noticed	
" No. 5,.....	" 16, 7 A. M.	29.175	Not noticed.	" 16, 8 A. M.	29.200	"	
Descending Nerqua,...	" 16, 10 h. 20 m.	29.300	88°	" 16, 11 h. 20 m.	29.300	88°	
" Ranch No. 6,	" 16, 4 P. M.	29.200		" 17, 7½ A. M.	29.365	77°	
				Mouth of Nerqua.			
" Nerqua, ...	" 17, 7 A. M.	29.275	75°	" 17, 8.13 P.M.	29.300	85°	
Mouth of Nerqua, ...	" 17, 8.45 "	29.345	85°				
First Truando Rapid..	" 17, 10.45 "	29.365	82°	Bottom of Rapids.			
Rancho No. 7,........	" 17, 3.30 P.M.	29.225	84°	" 17, 12 h. 25 m.	29.375		
				" 17, 7 P. M.	29.225	78°	Measurement of rain fall by Aneroid.
" No. 7,........	" 18, 7.30 A. M.	29.250	78°	Top of Main Fall.			
				" 18, 10 A. M.	29.312		
				Bottom of do.			
				" 18,	29.337		
Truando Lower Rapid,	" 18, 3.55 P.M.	29.275		" 19, 1.30.	29.265		
Rancho No. 8,........	" 19, 6 A. M.	29.250		" 20, 8 A. M.	29.300		
" No. 8,........	" 19, 8.4 P.M.	29.300		" 20, 8 "	29.300		
" No. 8,........	" 19, 2 "	29.400		" 20, 6 P. M.	29.300		
" No. 8,........	" 20, 7½ "	29.365					
Atrato R., Sucio,.....	" 22, "	29.375					
Quibdo,..............	MARCH 16, 10 A. M.	29.278	80°	MARCH 17, 10 A. M.	29.275	80°	
"	" 20, 10 "	29.324	80°	" 21, 4 P. M.	29.175	85°	
"	" 31, 10 "	29.275	79°	APRIL 3, 10 A. M.	29.265	80°	

TIDE TABLE—PARACUCHICHI.

Day of Month.	Time of High Water.	Height of Tide.	Phases of the Moon.
1855.	h. m.	Ft. in.	
January 17,	3.30 A. M.	10 11	New moon, 17d. 20.
" 19,	4.00 P. M.	11 8	
" 20,	6.30 "	12 6	
" 21,	6.30 "	12 2	
" 22,	7.15 "	11 9	

Highest tide, three days after change.

CALCULATED CUTTINGS, &C.

SECTION A.

From Kelley's Inlet to Rancho No. 1.

ROCK CUTTING.				SOIL.		
Length, Yards.	Depth, Ft.	Area of X Section, Sq. Yds.	Cubic Contents, Cub. Yards.	Breadth, Ft.	Depth, Ft.	Cubic Contents, Cub. Yds.
3,190	33	596	2,075,095	210	10	735,194
2,200	37	705	1,754,500	215	10	525,800
	47	890		215	10	
			3,829,595			1,260,994

Section A carried forward.

SECTION B.

From Rancho No. 1 to Rancho No. 2.

	ROCK CUTTING.						SOIL.		
Constant Section, 861 yds. Depth, 45 ft.	Length Yds.	Depth. Ft.	Depth Const. Sect'n. Ft.	Mean Br'dth. Ft.	Area of Whole X Sect., Sq. Yds.	Cubic Contents, Cub. Yds.	Br'dth, Ft.	D'pth Ft.	Cubic Contents, Cub. Yds.
	1,000	49	4	200½	950	1,022,500	213	8	189,333
	2,730	55½	10½	201	1,095	5,084,625	227	8	548,851
		121	76	209½	2,630				
						6,107,125			738,184

Section B carried forward.

SECTION C.

From Rancho No. 2 to Mouth of Tunnel.

		ROCK CUTTING.					SOIL.		
	Length Yds.	Depth, Ft.	Depth Const. Sect'n, Ft.	Mean Br'dth, Ft.	Area of Whole X Sect., Sq. Yds.	Cubic Contents. Cub. Yds.	Br'dth, Ft.	D'pth Ft.	Cubic Contents, Cub. Yds.
Constant Section, 846 yds. Depth, 44 feet.	1,280	124	80	210	2,712	3,897,600	228	5	162,134
	407	151	107	213	3,378	1,374,846	236	5	53,317
	974	151	107	213	3,378	3,650,069	239	5	129,321
	1,747	180	136	216½	4,117	7,487,642	244	5	235,845
		193	149	218	4,455				
						16,410,157			580,617

Section C carried forward.

SECTION D.

TUNNEL.

Central Section—One Tunnel.

Feet.	Areas, Ft.	
78×92.9×30 / 2	2,561	Low Water Section.
92.9″×12.4	1,144	To Level of Roadway.
39.9×102.9	4,084	To Radii of Arch.
Arch Radii: 63′ 6″ L 128½°	4,522	
9.	12,311	
	1,368	Sq. Yards×6,470 Yds.=8,850,960 Or in both Tunnels,..........17,701,920

Section D carried forward.

SECTION E.

From Rancho No. 5 to Mouth of Tunnel.

		ROCK CUTTING.					SOIL.	
Length, Yds.	Depth, Ft.	Depth Constant Section, Ft.	Mean Breadth, Ft.	Area of Whole X Section, Sq. Yds.	Cubic Contents. Cub. Yds.	Br'dth, Ft.	D'pth Ft.	Cubic Contents, Cub. Yds.
662	233	191¼	224	5,560	2,757,230	Nil.	Nil.	
	126	84¼	210½	2,770				
		Constant Section, 800 yards. Depth, 41¾ feet.						

Section E carried forward.

SECTION F.

From Rancho No. 5 to Rancho No. 6.

		ROCK CUTTING.					SOIL.	
Length, Yds.	Depth, Ft.	Depth Constant Section, Ft.	Mean Breadth, Ft.	Area of Whole X Section, Sq. Yds.	Cubic Contents, Cub. Yds.	Br'dth, Ft.	D'pth Ft.	Cubic Contents, Cub. Yds.
5,525	126	84½	210½	2,771	14,845,675	232	7	996,952
	119	77½	210	2,603⅛				
		Constant Section, 795 yards. Depth, 41½ feet.						

Section F carried forward.

SECTION G.

From Rancho No. 6 to confluence of Nerqua and Truando.

		ROCK CUTTING.					SOIL.	
Length, Yds.	Depth. Ft.	Depth Constant Section, Ft.	Mean Breadth, Ft.	Area of Whole X Section, Sq. Yds.	Cubic Contents, Cub. Yds.	Br'dth, Ft.	D'pth Ft.	Cubic Contents, Cub. Yds.
6,075	119	79	210	2,608	15,324,187	232	8	1,252,800
		Constant Section, 765 yards. Depth, 40 feet.						

Section G carried forward.

SECTION H.

From the confluence of Rios Nerqua and Truando to 1st Salto.

	ROCK CUTTING.					SOIL.		
Length, Yds.	Depth, ¾ Ft.	Depth Constant Section, Ft.	Average Breadth, Ft.	Area of Whole ✕ Section, Sq. Yds.	Cubic Contents, Cub. Yds.	Br'dth, Ft.	Depth. Ft.	Cubic Contents, Cub. Yds.
3,333	112	73	209	2,440	8,015,865	230	7	596,607
	109	70	209	2,370				

Section H carried forward.

SECTION I.

From 1st Salto to Rancho No. 8.

	ROCK CUTTING.					SOIL.		
Length. Yds.	Depth. Ft.	Depth Constant Section. Ft.	Average Breadth. Ft.	Area of Whole ✕ Section. Sq. Yds.	Cubic Contents. Cubic Yds.	Breadth. Ft.	Depth. Ft.	Cubic Contents Cub.Yds.
1,130	122	84	210½	2,689	2,957,210	Nil.		
1,716	116	78	210	2,545	3,851,562	Nil.		
3,410	91	53	207	1,944	5,865,200	Nil.		
1,645	72	34	204	1,496	2,048,025	Nil.		
	50	12	202	994				
					14,721,997			

Section I carried forward.

SECTION K.

From Rancho No. 8.

	Depths.		Section.		Mean Section.		Quantity.	
Lengths. Yds.	Alluvian Ft.	Rock. Ft.	Alluvian Sq. Yds.	Rock. Sq. Yds.	Alluvian Sq. Yds.	Rock. Sq. Yds.	Alluvian. Cubic Yds.	Rock. Cubic Yds.
1,416				776				1,098,816
4,248			564				2,395,872	

Section K carried forward.

SECTION L.

Lengths. Yds.	Depths.		Section.		Mean Section.		Quantity.	
	Alluvian Ft.	Rock. Ft.	Alluvian Sq. Yds.	Rock. Sq. Yds.	Alluvian Sq. Yds.	Rock. Sq. Yds.	Alluvian. Cubic Yds.	Rock. Cubic Yds.
1,970	29		815		822		1,619,340	
3,266	31		829		844½		2,758,137	
1,550	35½		860		829		1,284,950	
1,290	26½		798		805		1,038,450	
3,085	28½		812		778½		2,401,672	
1,812	19		745		790½		1,432,386	
	32		836					
							10,534,935	

Section L carried forward.

SECTION M.

To Rancho No. 9.

Length. Yds.	Depths.		Section.		Mean Section		Quantity.	
	Alluvian Ft.	Rock. Ft.	Alluvian Sq. Yds.	Rock. Sq. Yds.	Alluvian Sq. Yds.	Rock. Sq. Yds.	Alluvian. Cubic Yds.	Rock. Cubic Yds.
2,112	32		789		749½		1,582,944	
396	27½		710		697		276,012	
625	24		684		673		420,625	
1,312	21		662		678½		890,192	
2,580	25½		695		684		1,764,720	
978	22½		673		661½		646,947	
1,772	19¼		650		634½		1,124,334	
920	15		619		639		587,880	
1,208	21		659		664½		802,716	
250	22		670		655½		163,875	
2,089	18½		641		605½		1,264,889	
							9,525,134	

Section M carried forward.

SECTION N.

To Summit Level.

Length. Yds.	Depths. Alluvian Ft.	Depths. Rock. Ft.	Section. Alluvian Sq. Yds.	Section. Rock. Sq. Yds.	Mean Section. Alluvian Sq. Yds.	Mean Section. Rock. Sq. Yds.	Quantity. Alluvian. Cubic Yds.	Quantity. Rock. Cubic Yds.
4,630	15½		570		579½		2,683,085	
2,325	20½		589		556½		1,293,862	
308	14		524		561½		172,942	
2,825	21½		599		593½		1,675,225	
2,142	20½		587		603		1,291,626	
6,530	23½		619		556½		3,633,945	
3,334	11		494		499		1,663,666	
1,205	12		504		452		544,660	
	9		400				12,959,011	

Section N carried forward.

SECTION O.

To Atrato.

Length. Yds.	Depths. Alluvian Ft.	Depths. Rock. Ft.	Section. Alluvian Sq. Yds.	Section. Rock. Sq. Yds.	Mean Section. Alluvian Sq. Yds.	Mean Section. Rock. Sq. Yds.	Quantity. Alluvian. Cubic Yds.	Quantity. Rock. Cubic Yds.
2,560	9		400		390		998,400	
2,350	10		380		346½		814,275	
650	4		313		330		214,500	
2,920	7		347		358		1,045,360	
1,800	9		369		394		709,200	
2,464	13½		419		344		847,616	
484	0		269		294		142,296	
			319				4,771,647	

Section O carried forward.

RECAPITULATION.

	ALLUVIAN.	ROCK.	TOTAL.
Section A,	1,260,994	3,829,595	5,090,589
" B,	738,184	6,107,125	6,845,309
" C,	580,617	16,410,157	16,990,774
Extra Cutting mouth of Tunnel,		62,460	62,460
Tunnel,		17,701,920	17,701,920
Extra Cutting of Tunnel,		73,260	73,260
Section E,		2,757,230	2,757,230
" F,	996,952	14,845,675	15,842,627
" G,	1,252,800	15,324,187	16,576,987
" H,	596,607	8,015,865	8,612,472
" I,		14,721,997	14,721,997
" K,	2,395,872	1,098,816	3,494,688
" L,	10,534,935		10,534,935
" M,	9,525,134		9,525,134
" N,	12,959,011		12,959,011
" O,	4,771,647		4,771,647
	45,612,753	100,948,287	146,561,040

WILLIAM KENNISH, *Chief Engineer.*

Translation of a paper by a celebrated New Granadian engineer, given by his Excellency Nicomedes Conto, Governor of the province of Choco, to Mr. Kennish:—

Quibdó is situated in latitude 5° 36' 42" N., longitude 2° 39' 16" W., of the meridian of Bogotá, at an altitude of 42.8 metres above the level of the sea. Mean temperature 29° of the Centigrade; the maximum of heat 32°, minimum 26°. The hydrometer of Sauré makes the average humidity of the atmosphere 95°. From 90 to 100 inches of rain fall annually. The sea is distant, in a straight line, 145 maritime miles, or say 50 Granadian leagues, that is, from the extreme point of the Gulf of Urabá to the Bay of Candelaria, 55 leagues. The length of the river Atrato, from Quibdó to the Bay of Candelaria, is 91 leagues.

DISTANCES.

12 leagues to Beté.
20 " Bebará.
37 " Tebada.
38 " Fuerte Murrí.
42 " Mouth of Capica.
59 leagues to Murindó.
98 " Turbó.
10 " Lloró.
19 " San Pablo.

The highest part of the Isthmus of San Pablo has an elevation of 100.3 metres. Height of Tambo on Santa Helena, looking to the north, is 71.4 metres (the metre is 39 English inches.) Height of San Pablo, looking to the south, 60.7 metres. Width of Isthmus of San Pablo, 1 league, 1,000 varas.

Lloró lies in latitude 4° 22' 40" north, longitude 2° 34' 22" west of Bogotá, at an elevation of 69.1 metres above the level of the sea.

Urrao is situated almost due north of Quibdó; and the ravine or passage, between Farrallones and Cerro Platiado, is almost in a line with Quibdó. From Quibdó, a good road could be made, by following the line running from the head of Cabí to the Capon Hill, crossing the river Bebaramá at its sources, in order to strike the road from the heads of Titiribí.

CONFIRMATORY REPORT OF E. W. SERRELL, ESQ., CONSULTING ENGINEER.

NEW-YORK, 23d *August*, 1855.

TO FREDERICK M. KELLEY, ESQ., OF NEW-YORK.

DEAR SIR:

Agreeably to your request, I have examined with great care the general subject of connecting the Atlantic and Pacific Oceans, by means of water communication within the tropics; more particularly, however, upon the routes that have been surveyed, at your instance, during the past four years.

The task has been undertaken while appreciating its magnitude, and acknowledging an inability to master it properly.

Vast as the enterprise is, bearing directly or indirectly upon the political, commercial, social and moral relations of the civilized world—destined, when accomplished, to reduce space, by shortening the time of transit to and from the ends of the earth—the field of research, however, occupies comparatively a limited extent.

The excessive colds of the high northern latitudes, causing the uninterrupted presence of great masses of ice, together with the ever prevailing storms of a southern passage, long ago directed men's minds to a route through a more genial region.

Thus, for more than three centuries, almost superhuman exertions have been made by the best and most enlightened in their generation, to track out a path through the dividing slopes which separate the two great valleys of our planet.

Since the configuration, geographically, of the Western Continent has been known, but little reflection was required to determine that any water communication that should be established artificially, must be made within the parallels of 3 deg. south, and 20 deg. north latitude.

Within these limits many positions, at various times, have been indicated for making the desired connection.

The most prominent points, however, are as follows:—

Tehuantepec Route.

Commencing at the north with the route via. Tehuantepec, upon which, although it never was seriously contemplated to connect the oceans at this point by water, many very interesting explorations and surveys have been made, one of which is recorded by J. J. Williams, Esq., Principal Assistant Engineer to Major J. G. Barnard, who conducted an examination for the Tehuantepec R. R. Co., of New Orleans.

These explorations were very ably managed, and a vast amount of information was collected respecting the topography, geology, climate, local geography, productive industry, furnia and flora of the region.

The distance from ocean to ocean, by this route—from the sea near Tehuantepec, on the Pacific side, to Coatazacoalcos, on the Atlantic—is 186 miles. The summit at Nisi Correjor pass, is 855 feet high, and the general configuration of the country is a great gradual slope from either side towards the centre. The summit is more towards the Pacific side, than central.

No water exists in sufficient quantities, at the requisite level, to render this route at all practicable for a canal by lockages, and the enormous depth and length of any prism that could be formed by a thorough cut, precludes the possibility, commercially, of such an undertaking.

It is proper to remark that, although the explorations elicited other important and valuable information, they were undertaken with reference to a railway, and not water communication.

The direction of this route is nearly a south course from the Bay of Vera Cruz to the Pacific. The latitude is between 16 1-2 deg. and 18 1-2 deg. north, and the longitude is about 21 deg. west of Washington.

Many other surveys have been made over this Isthmus; that of Señor Moro justly deserves consideration—the essential features, however, are all given by Mr. Williams.

Honduras Route.

The Honduras Route, explored under the direction of E. G. Squiers, Esq., in the year 1853, lies eastward of the Peninsulas of Yucatan and Balize.

The direction of this line is also nearly south.

It commences at Puerto Cabello, in the Caribbean Sea, and terminates at the Bay of Fonseca on the Pacific.

The length of the route is one hundred and sixty statute miles. It lies wholly within the State of Honduras.

Here the valleys of the Humuya and Goascoran and the plain of Comayagua constitute a transverse depression from sea to sea, through the mountain range of the Cordilleras.

Here, also, as with the Tehuantepec route, however favorable the line may be for railway connection, it is utterly impracticable for adequate lockage for an inter-oceanic transit by canal.

Although the waters of the Rio Humuya, Rio Ulua, and Rio Goascoran interlock, and even pass by each other several miles on their course, from the interior to the great oceans, in either direction, the necessary summit water does not exist, by which lockages could be formed for a trade of any considerable extent.

The summit on this line, ascertained by the sum and length of the gradients given by Mr. Squiers, is 2,681 feet in elevation above the two oceans, and it is situated, topographically, somewhat similar to that on the Tehuantepec, that is, south-westward of the centre.

The latitude and longitude of this line, as determined by Mr. Squiers, is as follows: Puerto Cabello, latitude 15 deg. 49 min. N., and long. 87 deg. 57 min. west of Greenwich; Bay of Fonseca, on the Pacific, 13 deg. 21 min. N., and long. 87 deg. 35 min. west of Greenwich.

Good harbors are said to exist at the termini of both these routes, preference being given to one or the other, by different writers.

The next important line of communication between the two oceans, occurs by way of the Lake Nicaragua.

Several very trustworthy surveys have, at various times, been made of this route, undertaken at the instance of different or the same interests. The description of it here given is obtained from Col. Childs' Report, made to the American, Atlantic and Pacific Ship Canal Company, in the year 1852.

This line commences in the Bay of San Juan de Nicaragua, on the Atlantic, in lat. 11 deg. N., and long. 83 deg. 45 min. west of Greenwich, and runs in a north-westerly direction across the Continent.

Starting at the harbor of San Juan de Nicaragua, near the village of San Juan, or Greytown, it crosses several lagoons, and entering the valley of the river San Juan de Nicaragua, crosses the river at Juanillo, and ascending either in the bed of the stream, through slack-water navigation, produced by locks and dams, or by open cuts, through an independent prism, it was proposed to reach the summit by fourteen lockages, in a distance of one hundred and nineteen miles.

This summit level is the Lake Nicaragua, and the surface, at ordinary stages of water, is 105.24 feet above low tide in the Atlantic.

From the Lake Nicaragua, after cutting through a considerable ridge, the descent was proposed to be made to the Pacific at Brito, by another flight of fourteen locks, in a little over nine miles.

The entire length of this route, from the Atlantic to the Pacific, is as follows:—from the harbor of San Juan de Nicaragua to Port San Carlos, one hundred and nineteen miles; from San Carlos to Rio de Lajas, fifty-six and one-half miles; from Rio de Lajas to Brito, eighteen and one-half miles—making the whole distance one hundred and ninety-four miles.

This transit, for many years, has been considered very desirable, in consequence of the abundant supply of water on the summit, in the Lake Nicaragua.

Col. Childs, in his Report, suggested a canal having a prism fifty feet wide in the bottom, seventy-eight feet in the top, in rock cuttings, and one hundred and eighteen feet wide, in earth cuttings and lines of embankment, and having seventeen feet depth of water the entire distance from ocean to ocean.

It was proposed that the locks should be two hundred and fifty feet long, and sixty feet wide—average lift of lock, eight feet.

An artificial harbor of thirty-three acres was contemplated, at the Pacific, seventeen feet deep at low tide, and formed almost entirely by excavation in solid material, at the mouth of the Rio Grande.

Although comparatively of such small dimensions, and being intended for the transit of but one vessel at a time, except at the turn-outs, and that vessel only drawing less than seventeen feet of water, the estimated cost of this work, as computed by Col. Childs and his assistants, Messrs. Fay and Richmond, was thirty-one million, five hundred and thirty-eight thousand, three hundred and nineteen dollars and fifty-five cents. ($31,538,319 55c.)

Mr. James Walker and Col. Ed. Aldrich, English engineers of eminence, who were designated by the British Government, at the request of the parties in interest, in the main, corroborated the views as expressed by Col. Childs, in respect to the cost of the work.

The ability of this work, as proposed by Col. Childs, having such a depth of water, width of prism, width and length of lock-chamber, &c., to accommodate that portion of the trade of the world, for which any inter-oceanic canal might be intended, is hereafter referred to.

South-east of Nicaragua is the celebrated route of the Isthmus of Panama.

This line for ages was the favorite transit between the oceans; and having been known to Pizarro and Cortes, became afterwards the highway of the buccaneers. For the seventy years preceding the discovery of gold in the playas of California, at the close of the war between Mexico and the United States, it occupied almost exclusively all the attention that the civilized nations of the world devoted to an inter-oceanic junction.

Expeditions were fitted out at various times by the Spanish, French, English, Dutch, and Portuguese governments, to explore this wonderful and then almost fabulous region.

The most reliable and important of these expeditions were conducted, on the part of the French, by Napoleon Garilla with a corps of sappers and miners, and by Major Lloyd, an eminent English engineer.

The former of these gentlemen made a very elaborate report on the subject, contemplating a connection by railway and canal; and a grant was obtained from the New Granadian Government, authorizing the construction of a work of either character, and fixing the time for its commencement within a limited period.

The very great estimated cost, however, put upon this work precluded its adoption in either form, as contemplated by M. Garilla, and the period fixed for its commencement went by without anything being done.

Major Lloyd's triangulations and hydrographical observations added a great deal of information upon this interesting subject, and won for himself lasting renown; but it was not until our own lamented Stephens, (returning from the travels in Central America, Chiapas and Yucatan, which filled the world with his name,) entered boldly and determinedly upon the project of connecting the bay of Panama with a point near the mouth of the Chagres River by iron rail.

Some observations, made by James L. Baldwin, Esq., at Mr. Stephens' instance, settled a disputed point in reference to the gradients on the Pacific slope; and an expedition was fitted out in 1848, by a company of New-York merchants, of which Wm. H. Aspinwall, Esq., took an active and prominent part, to attempt a location for a railroad.

This expedition was headed by Brevet Lieut. Col. Geo. W. Hughes, U. S. Topographical Engineer. Messrs. Wm. H. Sidwell, Wm. Norris, Lloyd Tilman, Edward W. Serrell, James L. Baldwin, John May and J. J. Williams, conducted the surveys; and a location was made, upon which general line the present Panama Railroad has since been constructed, under the direction of Colonel Totten and Mr. John C. Trautwine, of Philadelphia, as principal engineers.

The length of this route, from the city of Aspinwall, on Manzanilla Island, in the bay of Limon, on the Atlantic, to the city of Panama, on the Pacific, is forty-seven and a half miles.

Commencing at the Atlantic, passing inland, the line traverses a swampy region, on either side of the Gatun River, which it crosses; thence follow-

g the base of the ridge Boho Soldado, in the valley of the Chagres, which
tter it crosses a few miles below Gorgona; thence takes the high ground,
rough the valley of the Rio Obispo, it reaches the summit at an elevation of two hundred and fifty-seven feet; thence descending in the valley
" the Rio Grande, crossing several unimportant streams, it enters
anama.

The very considerable length through the base of the water-shed of this
untry, together with the fact that no adequate sources exist for the supply of a summit level and lockages, if contemplated, must for ever preclude
the possibility of connecting the two oceans at this point by water communication, artificially, though a work of commercial value, notwithstanding the very great facilities that exist for maintaining a railway.

South of Panama, a great many explorations and surveys have been
made.

Besides those conducted by Col. Hughes in the country of the San
Blas or Mandingo Indians, Dr. Cullen, Mr. Gisborne, Capt. Prevost, of H.
I. S. S. Virago, Mr. Kennish, Lieut. Strain, U. S. N., Col. Codazzi, Capt.
oureiguiberry, and others at the head of private expeditions, or parties
ent out by the governments of New Granada, France, England and the
United States, whose researches, from time to time, have been made public,
have thrown more or less light upon the possibility of connecting the two
ceans by water communication through the territory lying between the
Atrato River and the Panama Railroad.

These researches have demonstrated that the Sierras Lloranas, or coast
ange on the Atlantic, extends from the westward of the gulf of San Blas,
n longitude about 79 deg. 30 min. west of Greenwich, in an unbroken
chain, to the Gulf of Uraba, or Darien.

It has been thought that depressions existed in the range, through
which an open cut, without locks, from ocean to ocean, might be made;
but to this time nothing of the sort has been shown to be practicable.
On the contrary, the expedition under Commander Prevost, which ascended the valley of the Savana, in 1853, from the Pacific, from the Gulf of
San Miguel, in the hope of reaching a very low pass to the Atlantic,
after attaining an altitude which they estimated at about twelve hundred
feet, and discovering no indications which confirmed them in the belief
that they had reached the summit, turned back to the Pacific.

Similar attempts were made about the same time by the parties headed
by Colonel Codazzi and Captain Joureiguiberry, from the Atlantic coast
at various points, and resulted equally unsatisfactorily.

The expedition headed by the brave, though unfortunate, Lieut. Strain,
U. S. N., after traversing the valleys of the Savana and Chucunaque for
several weeks, terminated fatally to a number of his party, but without
demonstrating even the probability of the existence of any passes through
which the desired communication could be made.

It may, therefore, be concluded that it is not within the limits of probability that a feasible line should be found, between the most westerly
mouth of the Atrato River and Porto Bello, notwithstanding the comparative narrowness of the Isthmus.

Entering the broad valley of the Atrato on the 77th degree of west
longitude, a new field of research is presented. It is to this region that
particular attention is now directed.

South of this point, although the valley of the Magdalena, the lake

Maracaybo, the Orinoco, and the great Amazon, all occur within 10 deg. of latitude, all of which, with the exception of the lake Maracaybo, cross transversely, very nearly, the entire northern portion of the Continent of South America—and some of them, with their head waters, are navigable more than nineteen-twentieths of the distance which separates the two oceans—yet the Cordillera range presents an impassable barrier. Upon this range occur some of the highest summits of the globe.

The Atrato River, designated, nearly half a century ago, by the great Humboldt, as one of the most probable routes by which to connect the two oceans, has been very minutely explored. Upon it, and the river San Juan, Mr. John C. Trautwine made an elaborate examination, and set forth the result of his labors, in a report made to Messrs. Belknap, James, and other gentlemen, in November, 1852. He has also published more detailed statements of these explorations, which were made with direct reference to an inter-oceanic canal, in the Journal of the Franklin Institute, of Philadelphia.

James C. Lane, Esq., of New-York, also made very extended explorations, in the years 1853–4, from the mouths of the Atrato to the Isthmus of San Pablo, near the head waters of the Rio San Juan, including the tributaries of the Atrato; as did also Capt. William Kennish about the same time, and subsequently at your instance, whose labors are hereafter more fully referred to. These gentlemen, with their assistants, among whom were Dr. Halstead, Mr. Brown, Capt. Rude, and Mr. Nelson, agree in the general features of the district which they passed over.

The Atrato River empties into the Gulf of Darien, by nine mouths, called by the inhabitants of the country "Bocas," and named as follows: Tarena, Candelaria, Pabo, Matuntubo, Coquito, Coco-Grande, Pantana, Uraba, and Leon.

Navigation into these mouths is impeded by sand-bars, and about ten feet of water, at low tide, is all that is usually found in the deepest, while some are dry.

Within the bars the water deepens suddenly, which is also the case seaward; and sixteen to eighteen fathoms of water is found at a comparatively short distance from the shore. The harbor is represented as safe, and accessible in all weathers.

The several streams which form the mouths of the Atrato, are called Caños. They vary in depth; but the Caño Coquito and Barbacoas, which form a cross connection with the other mouths, have a depth of thirty feet within about two miles, or less, from the ocean.

From the Caños, seventy miles up the river, the least depth of water is represented at forty-seven feet. Above this point to Quibdo, a distance of two hundred and twenty miles from the mouth of the Atrato, the least depth of water is from eight to ten feet, according to the seasons.

The entire length of the Atrato is over three hundred miles, following the sinuosities of the river.

Above the Caños it varies in width, from one-quarter of a mile to a mile and a-half, and even two miles in some places, for sixty-five or seventy miles. Above this point, to Quibdo, its least width is from five hundred to nine hundred feet; while a greater part of the way it is broader, with the exception of one narrow pass, a few miles below Quibdo.

The valley of the Atrato is nearly $3°$ in length, or about 170 miles, and varies in width from $1°$ to $1\ 1\text{-}2°$.

The entire formation of the bottom of the valley is a great alluvial plain.

The banks of the river are in no case high. They are generally levees, formed from the sedimentary deposit brought down by the current, and precipitated.

The banks of the tributaries, in many instances, from the continuation of this process through long periods of years, have raised considerably above the surrounding country, and, in some cases, the surface of the water is above the land on either side. This is said to be true of some portions of the Pocodor, Napipi, Bojaya, Sucio, Truando, and the Leon.

In the vicinity of the Truando and Sucio, which were both ascended and explored by Mr. Lane and Captain Kennish in detail, vast lagoons, or plains covered with water and trees, were discovered. These lagoons, in many places, are from twenty to one hundred feet deep of water, with mud of unknown depth in the bottoms.

A dense overshadowing forest of enormous trees, of very great variety, is said to cover the whole country. Underbrush and impenetrable thickets are said to abound. No rock or stone of any kind, excepting small detached, water-worn pebbles, has been discovered in any of the bottom lands of the Atrato. The soil, however, is auriferous, and produces gold and platinum, which is washed in small quantities by the natives.

Coal mines are said to exist in the Antiochian range, which forms the eastern slope. No cultivation, or but little, has been attempted.

There is but little current in the Atrato River. It is said never to exceed two and a-half miles per hour in any part of it, below the confluence of the Napipi.

The elevation of the river at the mouth of the Truando, which is sixty-three miles from the Atlantic, is about fifteen feet, or from fifteen to seventeen feet above mean level of the oceans. It varies inconsiderably at any season of the year: one and one-half feet is the greatest range recorded by any of the explorers here referred to.

From the Truando to the Napipi, the ascent in the river is not quite so gradual as lower down, the mouth of the Napipi being from 42 to 45 feet in elevation above the mouth of the Atrato.

The river at Quibdo, two hundred and twenty miles from the Atlantic, is represented at seventy-five feet above the oceans.

Rain falls in great abundance throughout the whole valley of the Atrato; and although there are differences of opinion as to the exact quantity, there can be little doubt that as many inches descend here, in the course of a year, as in almost any other part of the globe—but few days, consecutively, ever pass without some showers. Excessive storms of thunder and lightning sometimes prevail, for a week or more at a time, without interruption.

Bright sunshine succeeds the storm, but the thermal changes are less considerable than in most other localities, a few degrees only, being the greatest range that ever occurs.

Respecting these general physical features, topographically, hydrographically, geologically, mineralogically, botanically and meteorologically, there are no essential differences of opinion between the observers who are here cited; and many others who have been referred to, whose names are not given, also agree with them.

According to the surveys of Capt. Kennish and Mr. Lane, the Truando is navigable from its mouth, upwards, about thirty-eight miles for craft drawing ten feet of water.

This river, as has already been remarked, has its sources on the Cordilleras of the Pacific Coast range, and after traversing a rocky bed, over several rapids and falls, for a distance of about thirteen miles, enters the low country and passes through swampy lagoons and marshes to its confluence with the Atrato.

The drainage of the Truando and tributaries is about four hundred square miles.

The banks of the stream are frequently overflowed, and are levees, in some cases, above the surrounding country.

The sources of the Truando, the Nerqua and Hingador, rise on the hills to the westward of the valley of the Atrato.

The Cordilleras, of which these hills are spurs, are at this point extremely low, and are very near the Pacific.

So much for the physical features of this country, generally.

The Line via the Atrato and Truando Rivers.

This route commences at the estuary of the Atrato, and, as designated by Capt. Kennish, it is proposed to enter the Caño Coquito, one of the mouths or Bocas.

Here, at low tide, there is about four feet of water at the deepest point on the bar. The water gradually deepens as the river is ascended, and, at about two miles, is thirty feet deep.

From here to the mouth of the Truando, the Atrato is nowhere less in depth than forty-seven feet, and most of the way it is much deeper.

It varies in width, from a quarter of a mile to two miles, as has been stated in the general physical description.

It would thus afford, when the bar at its mouth shall be removed, the most ample facilities for navigation, by the largest class of steam sea-going vessels and sailing ships.

The Truando forms an angle with the general direction of the valley of the Atrato, of about seventy degrees. Its confluence is about sixty-three miles from the Atlantic, by the meanderings of the Atrato.

The Truando is here taken as the line of the proposed channel, and followed for thirty-six miles. At its confluence with the Atrato there is a small bar, and the water is only eighteen feet deep. A few yards up, however, it deepens to thirty feet, and maintains this depth for a mile and a-half, or more. Above this, for thirty-eight miles, it has an average depth of about fourteen feet, and is ten feet deep at the shallowest point.

At thirty-six miles from its confluence with the Atrato, the line diverges into the valley of the Nerqua, which is followed on solid ground.

This part of the line, and all the remaining distance to the Pacific—twenty-six miles—is to be through solid rock cutting.

The line, from the valley of the Nerqua to the Pacific, passes over a summit 506 feet above mean tide in the oceans, which summit is about three and one-quarter miles through at its base. This it is proposed to tunnel.

From here to the Pacific it follows the valley of a small stream, and debouches, at the coast, at Kelley's Inlet.

The distinctive features of this route are:—

First.—That the oceans can here be united by canal, through an open cut, without locks, guard gates, or impediments of any kind.

Second.—That while it is one hundred and twenty-five miles from ocean to ocean, more than one-half of the distance is, by nature, ready for the passage of the largest class vessels.

Third.—That the remaining distance is either excavations of bars, or river bottoms, under water, or solid rock, with slight earth covering.

Fourth.—That excellent harbors exist at either terminus, requiring but little improvement to make them as desirable as any in the world.

Thus it is proposed to form an open cut from ocean to ocean, without locks.

Of the Means and Facilities of Construction.

Under this head is included,—
1*st.*—The materials for mechanical works and other structures.
2*d.*—The opportunity that exists for obtaining workmen.
3*d.*—The ability to maintain a supply of provisions, clothing, &c.
4*th.*—The sanitary condition of the country.
5*th.*—The organization of proper executive departments.

Materials for Mechanical Works.

But few mechanical works have to be erected in comparison to the magnitude of the undertaking as a whole.

The first work is the projected jetty or line of pilings on either side of the entrance to the mouth of the Atrato, in the Caño Coquito.

Mr. Kennish says abundant timber can be easily obtained, at a short distance, of good quality, and that stone may be had in abundance at the quarries of Turbo, and boated to the works across the bay.

The excavation of the bar, being in sand and mud deposit, can easily be done by any of the several kinds of dredging machines now in use.

From the mouth of the Atrato to the confluence of the Truando no work of any kind will be required.

On the Truando, no bridges or aqueducts are needed, and nothing is to be done but the deepening of the stream, and the construction of such works as may be found necessary to prevent the wash of sedimentary particles into the main prism by the small streams; for these purposes, abundance of material exists within a short distance.

Beyond the works on the Truando, an open cut in earth and mica-slate rock is contemplated, and here, up to Rancho No. 5, no mechanical work will be required.

At or near Rancho No. 5, the river Nerqua must be turned, and some mechanical work is wanted. But, according to the evidence, there is abundance of stone and timber of good quality.

At the great tunnel, proposed between Ranchos Nos. 3 and 5, some masonry may or may not be required. It is, however, stated, that good stone can probably be had within the line of the works, which is in basalt, if the same is wanted.

Timber for all kinds of temporary works abounds on this division.

From the mouth of the proposed tunnel, at Rancho No. 3, to the Pacific, no mechanical work is wanted, except at the Chupipi River, where good material is found.

At Kelley's Inlet, and from there to the ocean outside the breakers, some works must be built, such as guards, probably jetties and lighthouses, but the material can readily be found near Rancho No. 2.

In fact, on so long a line of works, it is extraordinary to find so nearly everything required upon the ground; nothing to be used in the works need be brought from a distance, but cement and the metals.

The Opportunity that exists for obtaining Workmen.

The natives of this region, and more particularly those that can be obtained from the interior of the province of Carthagena, are said to be faithful, industrious, sober and prudent men, inured by nature to the climate and food of the country, requiring but little clothing, and thriving on the plantain and wild pork that abound, and the rice of commerce, which is readily obtained.

Very careful inquiries have been made of a number of gentlemen, reliable, in respect to the resources of the neighboring provinces, and from their opinions the conclusion is formed that at least six thousand men can be counted upon as a constant supply, if required.

It is a very different thing from having to transport men from great distances, when they can be had in their native country.

The Jamaica negroes, too, will do well in this country, and may be relied upon in great numbers.

Germans, Irish, English miners, and Americans for the tunnel, and Chinese, may be imported and comfortably sustained and maintained, as will more fully be seen by reference to the other executive departments.*

It may, and probably will, be necessary to institute a regular system of emigration from the older countries of Europe and Asia; and as the difficulties of distant land carriage will not be felt here, as in other places where attempts have been made to bring together large foreign populations, success in so doing will be less questionable.

The healthiness of the country is, also, very much in favor of the work.

At New-York, about one thousand per day arrive from Europe alone, to settle in the United States; upwards of three hundred thousand have landed here in a year; while, probably, when the work at the canal is once organized, and sufficient numbers are on the ground, from five to six thousand per year will be sufficient to supply the places of those who may remove, or become disaffected, or die in the country.

The Ability to Maintain a Supply of Provisions, Clothing, &c.

Good harbors existing on either side, will admit an unlimited commerce to be carried on, to the coast.

From the shores of the Atlantic, steamboat navigation already exists to Rancho No. 8, on the Truando.

---

* Since writing the above, the chief engineer of the Panama Railroad has published a report, by which it appears that the mortality on the Isthmus is much less than has been supposed. It appears that only 293 white men died in five years, out of 6,000 that were constantly engaged on the work. The coolies fared worse; the Jamaica negroes and natives of the Isthmus, better.—*Copied from the New-York Times of 23d Aug.*, 1855.—E. W. S.

Supplies can be brought to the mouths of the Atrato from all parts of the world, and transhipped up the river, when the bars at the mouths are deepened, which could be in about two years—large vessels could carry their supplies all the way to the confluence of the Atrato and Truando.

On the Pacific slope, the facilities are very great.

The harbors being accessible for large vessels and all kinds of steamers, supplies could be brought from Chili, Peru, the Sandwich Islands, California and Oregon.

All the breadstuffs can be delivered here as cheap as at New-York or Liverpool, and many articles of consumption at less prices: certain things, such as some kinds of clothing and luxuries, would cost more, but the average would compare favorably with any port on the Atlantic.

The valley of the Nerqua and the Pacific slope, are said to be fine producing countries.

Plantains and nuts of various kinds grow wild in great profusion, and the opinion of those who have devoted time and attention to the subject, is (in which several agree), that the country between the Rancho No. 7 and the Pacific, will sustain as large a population as the same extent of country in the best parts of Cuba.

It is said to be particularly well adapted to raising cattle and hogs: the latter furnish the favorite meat of the natives.

The Sanitary Condition of the Country.

From the Atlantic to the high country on the Truando, the climate is uniform, but excessively moist.

The thermal range of a great many observations, by different persons, spread over several months, indicated fluctuations of only 15 deg.

The very frequent and heavy rains, however, are unfavorable; and owing to the overflow of the rivers, the country is almost uninhabited, and uninhabitable.

But few natives are found anywhere through this region, but it fortunately happens that, with the exception of subaqueous excavations, no work, but a small amount of piling, has to be done on any part of this division.

From the high ground to the Pacific, over the valley of the Nerqua, and the water-shed on the westward side, the climate is as salubrious as the soil is productive, and men from all parts of the globe can work here, and live as well and comfortably as in any part of the tropics.

Those who are compelled to labor on and about the excavating machines, on the rivers eastward of the high ground, and at the mouth of the Atrato, must be provided with arks, or floating houses, made comfortable.

The Executive Departments.

It is probable that the work can be carried on in the best and most satisfactory manner, by regular departments, or bureaux.

These should include a chief executive or president, with an executive staff.

The president to be the supreme head of the work, and to manage the whole general business. His staff should be the heads of the other departments.

An engineer department, consisting of chief engineer, one principal assistant engineer on the ground, and three other assistant engineers, heads of divisions, together with the necessary subordinate field and office assistants and draftsmen as may be necessary.

To the engineer department should be intrusted all the general and detailed plans of the work, according to the usual methods. The advice of the most able and eminent engineers in the world should be obtained on the general plan in particular, and such a system of construction should be devised, that the details may be executed in the best way, at the least cost, and in the shortest possible time. The burden of the work will fall on this department.

A medical staff should be organized, and proper hospitalary arrangements made for the sick.

The medical staff might be constituted similar to the corresponding department of an army in quarters.

In consequence of the very great number of men that must be employed to do the work in such a time as the wants of the trade of the world require, and as the climate is hot, although generally healthy, the medical deparment should be organized in the most efficient manner, so that prompt and ample relief may be afforded at all times, on any part of the line.

A paymaster's department will also be required.

This bureau should take cognizance of all the expenses of every kind and nature whatever, and having suitable disbursing officers, should regulate the entire fiscal system of the enterprise that relates to the construction.

A commissary department is also needed.

The supplies of such a number of men as must be engaged on this work, will at first tax the energies of an experienced merchant; but with the progress of the enterprise, facilities will be created, towns will spring up, and merchandise will be brought into the country by persons not in the organization; but, at first, very great efforts should be made to subdue the natural wilderness, and to bring under cultivation large districts in the vicinity, in order that the best system of supplies may be maintained at the least cost, and without importation.

It is said that the district through which the heaviest portion of the work is to be constructed, will, with proper cultivation, maintain a much larger population than will be required upon the line of the canal, and those not engaged upon it, such as the families and friends of some of the workmen, who will follow, and will, in a short time, form a civilized community.

The readiness with which provisions, clothing, &c., can be had on the line of the works, may be judged from the fact that most of the labor will be expended on about thirty miles in length—one end of which rests on the Pacific, and the other at the head of steamboat navigation, on the Truando River.

But little over six miles is, therefore, the average haul for supplies, which at first may be carried from either end to the centre, by mules, over the mountain roads, and, in course of time, good wagon roads, or even a railway or tram-road could be constructed.

A department of machinery and repairs will also be required.

Dépôts should be erected on the Pacific, and at suitable points along the line, and at the mouth of the Atrato, where the boats and machinery employed may be repaired, and where the business of obtaining the necessary materials for the work for the dwellings, stores and hospitals, can be conducted.

A quartermaster's department, charged with producing suitable and healthy habitations for the men, and generally to care for their comfort, seems very necessary.

Provision must also be made for supplying the work with laborers and mechanics. This, perhaps, can be done best by a bureau attached to the engineer department, or by a separate organization.

The facilities for obtaining men, have been heretofore referred to.

As the laws of the land provide for the judiciary, no police regulations are necessary on the part of the organization constructing the work.

It is believed that a system of contracting, in the usual way for works of less magnitude, would not here be judicious, as the profits in such a plan would be very great, and should be realized by the original organization, while the advantages of separating the responsibility (the only argument in favor of the contract system that could apply here), may be obtained by the proper execution of the functions of the different departments.

It by no means, however, follows that great advantages may not arise from contracting distinct portions of the work or supplies.

Of the Cost of the Canal.

The work at the mouth of the Atrato consists of piling, crib-work, and dredging. Only an approximate idea of the cost here can be arrived at, until some more definite surveys have been made, and fixed upon. By the data hereto annexed of the work of dredging machines, it may be safe to estimate the cost of the improvements at the mouth of the Atrato, at $50,800.

The excavations upon the Truando will depend very much on the facilities afforded by the natural condition of the bottom; but the following figures show that twenty-five machines, working for eight years, may probably accomplish the work. But as the material will, in some cases, have to be carried a long distance—say, 300 to 500 feet, and deposited, it is not safe to estimate less than an increase of 100 per cent. on this sum, although a liberal per-centage for contingencies and breakages has been made.

ESTIMATE

Of amount of alluvial deposit to be taken out of the bottom of the Truando, and the number of machines that will be required to do the work in eight years:—

QUANTITIES.

	Cubic Yards.
Section K	2,395,872
" L	10,534,935
" M	9,525,134
" N	12,959,011
" O	4,771,647
Total amount	40,186,599

Twenty-one dredging machines, working at the rate of 800 cubic yards per day, for eight years, of three hundred days each, would be sufficient to take out the above; but allowing for breakage, &c., 20 per cent., it would require twenty-five machines.

ESTIMATE

On the cost of Dredging Machines to be used on the work:—

Number of Machines on Truando.................... 25	
25 Machines at $10,000 each..................	$250,000
Interest on $250,000 at 7 per cent., for 8 years, amounts to.....	$140,000
Cost of working 25 Machines for 8 years, or 2,400 days, at $9 per day...	540,000
Total cost..	$680,000

These machines are worked by about five or six men, and employ steam power.

The above contemplates the destruction of the machines in the time of service, which does not necessarily follow as a consequence.

ESTIMATE

Of the amount of subaqueous excavation of the Cano Coquito of the Rio Atrato:—

Taking the average depth of cutting 15 feet 6 inches, and the length of cut at 19,200 feet, and the average width of cut at 200 feet, the amount of excavation is 2,204,444 cubic yards.

Number of cubic yards of excavation, 2,204,444.

Five dredging machines, or subaqueous excavators, working 300 days in a year, and taking out daily eight hundred cubic yards of material, each, would excavate the above in less than two years.

The bar here, however, might be excavated by jetties, as proposed by Mr. Kennish, and by opening only a small channel through the centre, and allowing the current of the river to remove the remainder into deeper water.

ESTIMATE

Of number of men required to work on the sections between Kelley's Inlet and Townsend's Junction, supposing each man to take out 400 cubic yards per year, for twelve years:—

AMOUNT OF EXCAVATION.

	Cubic Yards.
Section A...	5,108,589
" B...	6,845,309
" C...	16,990,777
Extra at Mouth of Tunnel............................	62,460
Tunnel...	17,701,920
Extra at Mouth of Tunnel............................	73,260
Section E...	2,757,230
" F...	15,842,627
" G...	16,576,987
" H...	8,612,472
" I...	14,721,997
Total Amount.................................	105,293,628

Allowing 4,800 yards for each man in twelve years, there would have to be 21,934 men employed.

On this basis, it may be safe to say that an establishment of 25,000 men can perform the work in twelve years, and that this number must be kept constant.

The cost of maintaining such a number will greatly depend upon the efficiency of the departments of the Executive; but if they can be procured and maintained at an average of one dollar per day for the laborers, with mechanics, foremen, &c., in proportion, the cost of rock work, in the character of material here to contend with, should not, in open cut, exceed seven-eighths of a dollar per cubic yard.

The tunnel will not cost near so much per cubic yard as one of a small section, owing to the proportion that exists between the heading, always costly, and the remainder to be excavated.

It is believed that about $2 per yard will do it, which would be, say—ten dollars per yard for the heading, and over $1 70 for the remainder in the foot wall.

The works at the confluence of the Truando and open cut, and the necessary waste weirs and overflows, and constructions to prevent sedimentary deposits being precipitated into the canal from the small streams, including the turning of the Nerqua and Chupipi, can only be approximately determined in the most general way. It is, however, believed that ample allowance has been made to cover contingencies.

The cost of the improvements necessary at Kelley's Inlet must be also uncertain, until further surveys have been made; but an approximate quantity has been fixed upon, believed sufficient.

The great bulk of the work being the rock excavation, a more satisfactory conclusion is arrived at relative to it.

The following figures show, as nearly as circumstances will at present permit,

A SUMMARY OF THE ESTIMATED COST OF THE CANAL, AND APPURTENANCES.

Works at the mouth of the Atrato	$50,800
Excavations under water, in Truando	1,360,000
Excavations at the confluence of the Truando and Canal, including coffer-dams and pumping	40,000
Excavations between confluence, as above, and Pacific, (excepting Tunnel.) The quantities are all called rock; the excess by these measures being allowed for grubbing and clearing at $7/8 per cubic yard	77,883,994
Tunnel at $2, being $10 for heading, &c.,	25,403,840
Harbor and Kelley's Inlet,	150,000
Light-houses,	35,000
Piers,	20,000
Depots, (Pacific),	50,000
Depots and hospitals on line,	35,000
Depot and Hospital, Townsend's Junction,	15,000
Depot at Turbo, and improvements necessary,	70,000
	$105,113,634

Amount brought forward,	$105,113,634
Executive department, 12 years,	180,000
Engineer department,	562,000
Medical department,	120,000
Pay department,	140,000
Commissary department,	260,000
Quartermaster's department,	150,000
Supplies for do.,	550,000
Twenty-five pumping and hoisting engines, for the work in the great cut,	1,250,000
Contingencies, 25 per cent.,	27,081,408
	$135,407,042
If the rock is estimated at $1, this would be with contingent allowance about, *extra*,	10,000,000
	$145,407,042

It is not pretended that the above is more than an approximate estimate. It is, however, believed to be ample. The dredging work is generally done for one-half what is here allowed, and rock work in this character of material for much less.

In order to show that the great number of men necessary can work to advantage, a very simple calculation will suffice.

The line of the heaviest work is spread over twenty-five miles, so that about one thousand men are allowed to a mile.

Now, if the excavations are worked in terraces, commencing a quarter of a mile apart, and are carried forward in regular lifts of five or six faces each, according to the nature of the particular pit and its stratification, only about forty or fifty men need work at a face, which, being two hundred feet wide, and from twenty to thirty feet high, with the cross cuts, will afford ample room for the number proposed. In fact, many more may be employed, if necessary.

Machinery for taking out the material, and delivering it on the spoil banks, on either side of the cut, is estimated for, as well as pumping engines for keeping the work free from water, which will run in from the surface. The tunnel must be driven from either end in the heading and the foot work. The heading can probably be worked, also, from at least one shaft, if it is found necessary, in order to do the work in time, and when once completed, the remainder will be similar to the other parts of the open cut, except that it may be worked both day and night, and in all weathers, which is in its favor.

It may be proper to remark, in connection with this estimate, that this prism is intended to pass the largest class vessels, abreast, and without locks.

That a saving of many millions of dollars may be effected by changing the plan, there is no doubt.

Thus, by constructing the canal with a width sufficient for one ship at a time, with frequent turn-outs, a saving of about one-half of the above estimate would be made, reducing the entire cost to say $72,703,521; and by introducing one lock on the Pacific side, the depth of cutting through the rock would be correspondingly reduced, ensuring thereby a still further material saving in the cost.

All this, however, would not produce a perfect work, and is not here estimated for.

CALCULATIONS

RELATIVE TO THE COMMERCIAL VALUE OF THE CANAL.

The freight paid at San Francisco on Goods from the Atlantic, in 1853 alone, being on only a part of the trade, as some freights were paid before shipping, was $9,911,432. At least *three-quarters* of this charge would be saved—$7,433,574—or the interest, at 6 per cent., on........................ $123,892,900

The foreign trade to San Francisco that would be benefited one-half saving of freight, paid at port in 1853, $1,840,650. One-half of which is $920,325, or a sum equal to the interest, at 6 per cent., on............................... $15,338,750

The United States Foreign Pacific Trade was carried on, in 1854, in 961 vessels bound out, and 895 entered home, showing that the trade is equal to over two-thirds that of our domestic California trade, and would therefore, by passing this Canal, make a corresponding saving, or a sum equal to the interest on... $99,450,570

Thus showing that the Canal would save to the United States alone, the interest on the sum of........................ $238,682,220

The trade of France with the Pacific, which would pass through the Canal, and be benefited about one-half, is a little over $30,000,000. Our foreign Pacific trade being a little over $33,000,000, the ratio of saving is in the direct proportion, or a sum equal to the interest, at 6 per cent., on............ $90,000,000

The British trade through this Canal would be, if no increase was estimated, $180,000,000; which, by the same rules of saving in comparison to the distance sailed, as govern the calculations relative to the United States, would be equal to the interest, at 6 per cent., on the sum of $540,000,000. But as the trade has facilities via the Cape of Good Hope, say to be benefited by Canal... $300,000,000

The statement would then be:—

United States,.. $238,682,220
France,.. 90,000,000
England,... 300,000,000
Other countries,... 100,000,000

$728,682,220
Six per cent. on... $200,000,000

for cost of Canal.
And the interest, at 6 per cent., on......................... $528,682,220
saving to the world.

Or, twelve per cent. profit on the cost of the Canal, at $200,000,000, and a benefit to the world at large of the saving of interest, at 6 per cent., on $528,682,220 annually.

The above figures, although made from the official returns, are only approximate, as they do not include many items of trade that now exist, and no allowance is made for the increase of trade, whatever.

The data is from Mr. Stone's Commercial Statements.

CONCLUSIONS.

It will be borne in mind that the conclusions here arrived at relative to the practicability of this enterprise, and, consequently, its commercial value, are based mainly on the data furnished by Mr. Kennish, partially confirmed and corroborated by Messrs. Lane and Trautwine.*

The field notes and surveys of Mr. Kennish have been very carefully examined in this connection, and it is only justice to him to say that they are consistent with themselves, and present the appearance of care and truthfulness.

Where so much depends on the accuracy of observation and faithfulness of record, too much care cannot be taken to guard against errors. In this case, however, there seems to be every probability that the surveys represent faithfully what they purport to, and may be taken for such, unhesitatingly.

It was not intended by them to represent the minutiæ of details.

In my opinion, no other route yet discovered, by which to connect the two great oceans by water, can at all bear any comparison with this.

The Honduras and Tehuantepec routes are impracticable for water intercommunication, and the Isthmus of Panama is not physically constituted to admit the construction of a Canal within the limits of commercial availability.

The Nicaragua route, although possessing water on the summit, to be made as large and convenient for the business as that proposed at the Atrato, would cost considerably more, and in many other respects would be less favorable, inasmuch as it is sixty-nine miles longer, and has twenty-eight locks upon it. A corresponding width and depth of prism would produce more quantities on this route than the Atrato route.

And it is no less remarkable than true, that the Atrato route was indicated by Baron Humboldt, the greatest of modern observers, when in the country, and that, although referred to by such high authority, was comparatively neglected until the subject was taken up by yourself.

Opinions have at times been expressed that ships could be taken from ocean to ocean by railway, and that this method would preclude the use of a canal.

There is, however, no foundation for such a position.

If the ships could be taken out of the water, carried on a railway fifty or one hundred miles without injury to them or their cargoes, and then returned to the water again in as good condition as when taken up, (all of which is very problematical,) the expenses of maintaining such a road, and the cost of working it, estimating only one-half cent per ton per mile for *power* and expenses, would exceed the interest on an investment

---

* But it is due to Mr. Kennish to say, that he is the first engineer who has ever proposed to unite the two oceans by a passage up one mouth and down another, of a never-failing river, without any locks, which is the feature that distinguishes this route from all others.

twice as great as here proposed for an open permanent canal, without any impediment, as free to pass through as the ocean, and safer.

It is worthy of especial notice that the mouth of the Truando is situated relatively to the Atrato and the two oceans, within a very short distance of where, theoretically, it would be best to leave the Atrato River to enter the Pacific Ocean.

If the tides on the Pacific coast were in all places alike, this remark would not apply; but as the tide rises much higher, and falls lower, above and below mean level, at the Gulf of San Miguel than at Kelley's Inlet, and as the rise in the Pacific along this entire coast is very much greater than along the correspondingly opposite Atlantic coast, in the proportion of about twelve to one, it is evident that if an open cut were made through the Isthmus at San Miguel Bay, the Pacific Ocean would flow through into the Atlantic at high water, and the Atlantic would pour into the cut at low water on the Pacific; thus producing such great tidal currents that the work would be useless unless protected at either end by guard locks.

While, if a connection should be attempted above the confluence of the Napipi with the Atrato, the difference in altitude between this point and the Pacific being about forty-five feet, and the distance about sixty miles, at least three locks would be required to descend to the Pacific; while in the present location, by the way of the Truando, the descent to the Pacific from the confluence with the Atrato, being about equal to the descent of the Atrato to the Atlantic, and the distance being essentially the same, the current on both sides of the summit would be about equal, excepting the effect of the rise of water in the Pacific, which will oscillate in the open cut, and flow to the lagoons near the summit level.

The very permanent character of the work proposed must commend it to every one; no destructible material being used, and no attention being required forever after, to keep it in order, other than what must from time to time be given to the bars at the harbors, should it be found necessary to dredge them.

The distances that will be saved to the trade of the world by this route, being presented approximately in Mr. Stone's commercial statements, now in your possession, no occasion exists to repeat them. They would otherwise be given.

While submitting this memoir, I acknowledge my obligations to Mr. Kennish for the frank and open manner in which he has replied to the most minute and searching questions propounded to him, respecting the surveys, and the country, and the facilities afforded for constructing the work, as well as to James C. Lane, Esq., who was in charge of one of the field parties on a former occasion, and to Horatio Allen, John B. Jervis and A. F. Edwards, Esq., eminent Civil Engineers, and Col. J. J. Abert, U. S. Topographical Engineer, from whom valuable suggestions have been received.

Wishing you and the great enterprise in which you are so nobly engaged, all success,

I remain, your ob't. serv't,

EDWARD W. SERRELL, *Civil Engineer.*

THE AQUEDUCT,

IN ITS INFLUENCE UPON THE COMMERCE OF THE WORLD.

Having thus shown the feasibility of the proposed route, we may be allowed to add a few statements bearing upon the importance of this enterprise, in its relation to the commerce of the world, kindly furnished by David M. Stone, Esq., one of the editors of the New-York *Journal of Commerce*.

The project for a ship canal at the Isthmus was originated many years ago, before the discovery of gold in California and Australia had given such importance to the commerce of the Pacific. Even the early Spanish adventurers sought for a practicable route, and almost every important state or government on this continent, and in Europe, has, at one time or another, entertained the hope of finding it. Any one who will cast his eye upon the Map of the world, will see at a glance by what a tedious and dangerous route the commerce of the East Indies, China, Japan, Australia, and the settlements on the Western coast of the American Continent, must reach the principal markets of Europe and the United States.

The extreme length of South America is four thousand six hundred miles, and at certain seasons of the year, the passage around Cape Horn is not only protracted, but highly dangerous. Most of the European vessels bound for the East Indies take the eastern route, and double the Cape of Good Hope; but this is also a long and dangerous voyage.

To show the saving of time which would be effected in voyages from New-York to the Pacific, by a ship canal at the Isthmus, we extract the following table, prepared for this purpose by Lieut. Maury, and embodied in a Report of a Select Committe of the House of Representatives, presented in Congress Feb. 20, 1849.

TABLE

Showing the saving of time from New-York, by the new route, via the Isthmus of Panama, as compared with the old routes, via Cape Horn and the Cape of Good Hope, to the places therein named, estimating the distance which a common trading ship will sail per day to be one hundred and ten miles, and calculating for the voyage out and home:—

	Distance via Cape of Good Hope. Miles.	Length of Passage Out and Home. Days.	Distance via Cape Horn. Miles.	Length of Passage Out and Home. Days.	Distance via the Isthmus of Panama. Miles.	Length of Passage Out and Home. Days.	Saving by the Isthmus over route via Cape of Good Hope. Days.	Saving by the Isthmus over route via Cape Horn. Days.
To Calcutta.....	17,500	318	23,000	418	13,400	244	74	174
" Canton.......	19,500	354	21,500	390	10,600	192	162	198
" Shanghae.....	20,000	362	22,000	400	10,400	188	174	212
" Valparaiso...			12,900	234	4,800	86		148
" Callao.......			13,500	244	3,500	62		182
" Guayaquil....			14,300	260	2,800	50		210
" Panama.......			16,000	290	2,000	36		254
" San Blas.....			17,800	322	3,800	68		254
" Mazatlan.....			18,000	326	4,000	72		254
" San Diego....			18,500	336	4,500	82		254
" San Francisco.			19,000	344	5,000	90		254

The immense saving of time by the Isthmus route, has attracted the attention of all interested in commerce, since the American continent was first discovered.

Cortes, about three centuries ago, searched in vain for a natural passage between the two oceans; and many others followed the same course.

When a canal was first proposed, English capitalists joined heartily in the scheme, and the shrewdest foreign merchants and bankers have been ever ready to listen to any suggestion for its accomplishment.

If this were considered important twenty years ago, how much more so now, when the trade between Great Britain and her Australian colonies, and other near ports, has increased from three millions of dollars to seventy-five millions dollars; when California has opened to the commerce of the world her golden gates, and Japan is ready to renew, on a grander scale, her share in the trade of the world.

The trade between California and Atlantic ports will soon be almost sufficient, of itself, to support such an enterprise as we propose. The total amount of freight money paid at San Francisco, for the year ending 1853, was $11,752,084, nearly all of which was for cargoes that went from northern ports, around Cape Horn. The following is a summary of these payments:—

Payments as Freight Money at the Port of San Francisco, during the year 1853.

	From Eastern Dom. Ports.	From Foreign Ports.	Total.
January	$981,773	$204,172	$1,185,945
February	611,609	188,756	800,365
March	927,670	254,079	1,181,749
April	727,203	191,174	918,377
May	811,728	139,077	950,805
June	776,069	147,394	923,463
July	1,440,341	78,668	1,519,009
August	998,917	188,813	1,187,730
September	894,649	105,185	999,834
October	797,575	127,672	925,247
November	506,311	159,267	665,578
December	437,587	56,395	493,982
Total	$9,911,432	$1,840,652	$11,752,084

The above only shows the freight money actually collected at the port of San Francisco. Vessels from Europe, and many from domestic ports, received a great portion of their freight money at the port of clearance, and this, of course, is not included. This is true, to a still greater extent, of the commerce of 1854; a large portion of the freight money not being collected at San Francisco, and yet the total thus collected in that year amounted to $5,311,032, as will appear from the following summary:—

Payments as Freight Money at the Port of San Francisco during the year 1854.

	Merchandise from New-York.	Boston.	Other Atlantic Ports.	Foreign Ports.	Total.
January	$482,458	$196,603	$17,466	$149,983	$846,510
February	238,336	321,896	32,480	164,861	757,573
March	160,181	120,180	40,364	152,347	473,072
April	193,776	141,200	10,000	43,981	388,957
May	195,433	135,693	9,350	50,913	391,389
June	134,000	109,581	26,611	104,503	374,695
July	137,986	103,485	3,801	32,441	277,713
August	221,536	38,861	39,703	122,837	422,937
September	210,716	146,754	3,000	75,037	435,507
October	202,103	65,248	46,670	36,711	350,732
November	146,985	96,201	15,623	42,628	301,437
December	149,327	67,317	......	73,866	290,510
Total	$2,472,837	$1,543,019	$245,068	$1,050,108	$5,311,032

The tonnage engaged in the California trade is very large, the arrivals at San Francisco in 1853, being 1,028 vessels, amounting to 558,755 tons, and the clearances 1,653 vessels, amounting to 640,072 tons. In 1854 the arrivals were 620 vessels of 406,114 tons, and the clearances 1,193 vessels, of 515,861 tons. Of the above, a very large proportion came from the Atlantic *via* Cape Horn.

But the trade of California, large as that is shown to be, is not the chief reliance of the friends of the proposed canal. The trade between the United States and foreign Pacific ports has been, in a great measure, overlooked by those not directly engaged in it, and its importance has never been sufficiently estimated. We have obtained from the Department at Washington some reliable information upon this subject, which will be found full of interest.

These statements show that the total of clearances from the United States to foreign Pacific ports, and arrivals from the same, have increased about two hundred per cent. since 1849 (the tonnage showing a still larger increase, owing to the larger vessels employed), while the total imports and exports have doubled during the same space of time.

This will appear from the following summary, taken from official returns.

Commerce between the United States and Foreign Pacific Ports.

Fiscal Year.	No. of Vessels.	Tonnage.	Value of Imports & Exports.
1849	626	244,035	$17,001,320
1850	1148	383,574	18,402,103
1851	2040	673,619	21,375,428
1852	1693	661,980	27,013,110
1853	1803	787,707	31,554,653
1854	1858	957,599	33,953,456

The above is totally exclusive of the commerce between the Atlantic ports and California, and presents an increase of trade truly astonishing. In order to show the materials from which the above is compiled, we annex the details for the last year mentioned.

Commerce between the United States and Foreign Pacific Ports, for the Year 1854.

	Entered from.		Cleared to.		Exports.	Imports.
	Vessels	Tons.	Vessels	Tons.		
Dutch East Indies	20	8,986	18	7,951	$184,776	$1,041,609
British East Indies	96	58,043	64	46,397	636,412	5,378,321
Australia	19	8,342	105	44,410	3,149,079	214,202
Manilla	37	24,549	20	17,641	74,502	2,965,282
Chili	107	41,719	112	44,774	2,193,259	3,332,167
Peru	278	179,722	251	158,510	685,155	1,005,406
China	107	70,426	121	87,205	1,398,088	10,506,329
Minor Pacific Ports	123	42,264	155	50,560	952,815	10,103
Do. Indian Ocean	6	1,352	6	3,814	200	60,730
Japan			1	500		
Sandwich Islands	73	23,738	69	21,252	55,891	119,130
Northwest Coast	29	6,490	39	8,954		
Total	895	465,631	961	491,968	$9,330,177	$24,633,279

Of this, 1856 vessels, amounting to 957,599 tons, and carrying $33,953,456, value of merchandise, it is evident that nearly every sail would pass through the Ship Canal at the Isthmus, if such a one were constructed. In further confirmation of this, we can see what preference was given tó the clippers over ordinary sailing vessels—time being reckoned of so much value. We can hardly conceive it possible for a ship, clearing from New-York for San Francisco, to prefer a voyage of 19,000 miles, when by the use of a Ship Canal, without danger or detention, it could shorten the distance to 5,000 miles!

The interest upon the ship and upon the value of the cargo, and the cost of victualing and sailing the vessel, to say nothing of the dangers incident to a protracted voyage around the southern point of the Continent, would determine every vessel making a voyage from any American port on the Atlantic to any port on the Pacific, to adopt the advantages offered by the canal.

The trade between the various European States and the Pacific is large, and much of it would, undoubtedly, pass through the canal, if such a channel of communication were completed.

The following will show the number of ships, the tonnage, and the imports and exports, between France and the Pacific ports, during the year 1852:—

	No. of Ships.	Tons.	Imports and Exports.
British India,......	54	18,250	45,200,000
Society Islands,....	81	23,666	40,300,000
Peru,.............	32	11,051	25,200,000
Chili,.............	7	2,178	19,300,000
French India,......	25	7,886	13,400,000
Dutch India,......	16	5,076	7,300,000
Philippine Islands,..	4	1,252	1,700,000
China,............	3	912	1,100,000
Bolivia,...........	1	500	100,000
Total,.........	223	70,771	fs. 153,600,000

A total of 223 vessels, 70,771 tons, and imports and exports amounting to 153,600,000 francs, or over $30,000,000, for the trade of a single year.

In turning to the trade of Great Britain, we hardly know where to draw a line as to the portion of her immense commerce likely to avail itself of the advantages of the new route. Her direct carrying business between her own East India possessions and the home ports, is nearly $100,000,000 per annum, while to that is to be added the China and other East India commerce. Her traffic between the ports in the United Kingdom and the Western coast of North and South America, would certainly seek the new channel; and this, reckoning the cargoes both ways, now amounts to upwards of $30,000,000. The most important branch of British trade *certain* to pass through the Canal, is that connected with Australia; and here, lest we might be accused of exaggeration, we propose to give the official figures.

The following are the declared value of Imports and Exports at the Colony of Victoria, for the years 1853 and 1854.

IMPORTS.

	1853.	1854.
Building Materials,	£2,004,793	£2,436,071
Dry Goods,	2,869,542	3,592,093
Iron and Iron Manufactures,	1,059,261	1,354,653
Leather and Leather Manufactures,	461,994	527,311
Provisions and Groceries,	1,460,245	1,701,203
Specie,	1,163,344	87,480
Spirits,	1,015,053	839,704
Sugar and Molasses,	486,614	427,061
Tea,	205,364	292,837
Tobacco,	357,396	194,423
Wine,	373,524	409,075
Miscellaneous,	4,385,502	5,858,396
Total Imports,	£15,842,632	£17,720,307

The following will show from what countries the above were imported:—

	1853.	1854.
Great Britain,	£8,088,226	£11,050,329
British West Indies,	14,973	20
British North America,	13,560	60,238
Other British Colonies,	5,036,311	4,394,936
United States,	1,668,606	994,692
Other Foreign States,	820,961	1,220,092
	£15,842,632	£17,720,307

EXPORTS FROM VICTORIA.

	1853.	1854.
Gold,	£8,644,529	£8,255,550
Wool,	1,651,871	1,618,114
Miscellaneous,	761,143	1,901,540
Total Exports,	£11,057,543	£11,775,204

The following will show to what countries the above were exported:—

	1853.	1854.
To Great Britain,	£9,875,624	£10,270,213
British Colonies,	942,741	1,378,107
United States,	19,646	50,933
Other Foreign States,	223,532	75,951
Total Exports,	£11,057,543	£11,775,204

These statements, showing a total of £39,495,511 imports and exports for a single year, would be almost incredible, were they not taken from the official returns.

We might go further, and give in detail the commerce of each of the South American States bordering on the Pacific, where trade has wakened into new life within the last two or three years, but the foregoing will serve the purpose of illustration.

These are not estimates of a commerce likely to be created by the construction of a canal, but statements of a commerce now in being, a large portion of which would avail itself at once of this inter-oceanic communication. With such a business before us, the contemplated cost of the canal sinks into comparative insignificance.

A very small tax upon such an immense amount of tonnage, would keep the works in repair, and pay a large interest upon the capital expended.

To say nothing of the lessened risk, and the consequent reduction in cost of insurance, the convenience to the world of such a speedy transit from ocean to ocean, without breaking bulk or unbending a sail, can hardly be over estimated.

The advantages to be gained by a vessel in passing through the canal, over the route around the Horn, are so apparent, that any shipowner or merchant can calculate them from the data above given. To those interested in this enterprise, who are not practically familiar with nautical matters, it may be well enough to give a simple statistical statement, by way of illustration. This is a utilitarian age, and many may ask—"Apart from the lessened risks of navigation, what will the saving of time by the Isthmus route be *worth in dollars and cents?*"

Lieut. Maury has given an estimate of the time, in a table included above, which covers the average of voyages by all classes of vessels, not including detentions. We propose to give a brief comparison with a trip by a fast sailing vessel, which recently made a prosperous voyage:—

		Miles.
Distance sailed from New-York to San Francisco		19,000
Distance *via* the Isthmus	5,000	
Add for the proposed new route	700	
		5,700
Saving of distance, miles,		13,300

The passage around the Horn was made in 150 days, and at this rate the passage by the proposed canal would be 45 days, making a saving of 105 days. The following were the monthly expenses of the ship:—

1 Captain, wages per month		$150
3 Mates, at $50, $30 and $25 per month		105
1 Cook, per month		25
1 Steward, per month		25
1 Carpenter, per month		30
30 Men, $18 per month		540
37 Officers and crew, costing per month		$875

Amount brought forward,...	$875
Cost of provisions, and finding, ($2 each per week)..................	333
Present value of ship $84,000, on which 7 per cent. interest, and 8½ per cent. insurance, in all 15½ per cent., would be per month..........	1,085
Wear and tear, and depreciation, 10 per cent per annum, would be, per month,..	700
Total cost of vessel per month............................	$2,993
This for 105 days would be...................................	$10,475

Thus, to say nothing of the *profits* of the ship, the saving to the vessel in this voyage, by the use of the canal, would have been $10,475, less the charges for toll imposed for the passage. In other words, if the vessel had paid ten thousand dollars for the use of the canal in a single passage, she would have been the gainer! But this is only the gain to the *ship* itself. The owner of the cargo would make a large saving, even supposing that he paid the same rate of freight, with a cargo of only $50,000, the saving of interest on the voyage would be $1,020 81, and there would be few cargoes by such a vessel which did not greatly exceed this amount. This is the saving on a single trip of one vessel. Multiply this by the number of ships engaged in the Pacific trade, and reckon it out and back, and then estimate by this data the profits of the proposed work!

The wrecks which strew the tempestuous coasts around the lower extremities of the two continents, or repose far beneath the surface of their heaving billows, will be left alone, without the annual additions made to their sad catalogues by fresh disasters.

ERRATA.

On page 26, third line from top, for "Explanation of Plates," read Explanation *of Section or* Plate 1.

On page 27, on ninth line from top, for "perpendicular lines," read perpendicular *dotted* lines.

On page 30, on the third and twenty-sixth lines from the top, for the word "Sugra," read *Tuyra*.

On page 32, on third line from top, read, in place of, "of either ocean," the words, *of the dividing ridge.*

On page 32, on nineteenth line from top, for "head winds," read *headlands.*

On page 43, in column of remarks, for the words, "Measurement of rain fall by aneroid," read *Measurement of main fall by aneroid barometer.*

Printed in Poland
by Amazon Fulfillment
Poland Sp. z o.o., Wrocław